Life and Spirit in the Quantum Field

Spirit is real, feelings rule and other adventures in quantum life

By Doug Bennett

Doug Bennett
200 W. Jordan St.
Brevard, NC 28712

Copyright © 2016 by Doug Bennett

All rights reserved. Except as permitted under the United States Copyright Act of 1976, no part of this publication in any format, electronic or physical, may be reproduced or distributed in any form or by any means, or stored in a database or retrieval system without the prior written permission of the publisher.

ISBN-13: 978-1535382106
ISBN-10: 1535382104

Editor: Pat Bennett
Cover, typesetting/graphic design:
Doug Bennett
Illustrations by Kat Fitzpatrick and Doug Bennett

Printed in the United States of America 1 2 3 4 5 6 7 8 9 10

Publication history
Published in 2010 as ISBN 978-0-9815818-1-1
Published in 2016 as ISBN 1535382104

Acknowledgements

Getting to this book has been a long journey. Along the way, many people have generously helped me. I would like to acknowledge all the people who did the research and wrote the books I read and learned from. For those who have taught, or even tried to explain something to someone else, you know that teaching is the best learning vehicle there is. I appreciate all the students who invested their time with me, asked questions and shared their experiences.

My daughters, Kim and Cheryl, have been enthusiastic supporters of my efforts. My wife, Pat, has shared my interests and has always been willing to share a subtle nudge to keep me on task. She did a great job critiquing and proofing, too.

I have been invited to speak at the Haden Institute Summer Dream Conferences for nine years now. It was the support and encouragement I received from the staff and participants that led me to apply my mechanisms to issues of spirit. They have attended my workshops and bought my books. Alan Ballew provided thoughtful comments and questions. De Ila Meyer did a very thorough proofreading and markup. Kat Fitzpatrick looked at the illustrations I did in the preliminary draft and graciously offered to do real illustrations. They grace these pages. I am grateful.

Finally, I appreciate the knowledge and support provided by Kathleen Barnes, the book midwife at Take Charge Books.

Doug Bennett
August, 2016
Brevard, NC

Contents

Introduction 1

Part 1 Our Starting Place 9
Chapter 1 Building Reality 11

Chapter 2 Our Current Models and Paradigms 23

Chapter 3 Science and Spirit 45

Part 2 Applying Quantum Mechanics to Life 53
Chapter 4 Everywhere in the Hologram 55

Chapter 5 The Quantum Field 73

Chapter 6 Reality and Quantum Weirdness 85

Chapter 7 Holograms in the Brain 107

Chapter 8 Connecting to the Field 129

Chapter 9 Life Is Light 151

Chapter 10 Influence in the Quantum Field 175

Part 3: Quantum Life: The Possibilities 197
Chapter 11 Thinking in the Field 199

Chapter 12 Human Influence 219

Chapter 13 The Future 241

Part 4: All is One, Ye Are Gods 251
Chapter 14 Mechanisms of Spirit 253
Chapter 15 Spiritual Values 279

Part 5 Living in the New Reality 291
Chapter 16 The Sciences 293
Chapter 17 Religions 309
Chapter 18 People 321
Appendix 1 The Equations 335
Appendix 2 What Do I Do Now? 339
Index 349
About the Author 357

Introduction

WE HAVE A hard time with spirit in our culture. We have science that tells us that it is not real. We have religions that tell us that only heaven is important and the bodies and inner experiences we have are not important. And then we have the ongoing fight between science and religion. Through all this people continue to believe in the existence of a non-material world, where spirit and spiritual experiences are real, where knowing and action at a distance are possible and even common occurrences. It seems that what's real depends entirely on who you talk to.

My view is that when beliefs persist for a long time, it means there is something real underneath the beliefs, which may differ from the specific beliefs. This is a useful belief because much of modern science and much of organized religion suffer from incomplete views of reality, even though both have been pretty forceful promoting those limited views of reality as a complete view. The result is that our inner experiences are often different from what we are hearing from the big institutions in our culture. It is a conundrum.

I like to think that many of the old, persistent spiritual truths are true, especially those that appear across many cultures. Later in the book after much discussion of mechanisms and science, I will argue that two old spiritual aphorisms are literally true: All Is One (from Eastern spiritual traditions) and Ye are gods (from Psalms 82:6).

A note on capitalization of g/God: In my dictionary, god is defined as "any of various beings conceived of as supernatural . . ." And, God, is defined as "in monotheistic religions, the creator and ruler of the universe . . . " Capital G, God, then, refers to someone's specific god, while lower case, g, god refers to the general concept of god. Most of the time in this book I will be referring to the general concept, so I will use the lower case g. If I think I am referring to the god of a specific religion, I will use the upper case G.

Often, old wisdom, particularly spiritual wisdom, has a way of coming back as new truths. Consider, as an example, the popularity of the "All Is One" bumper stickers. That bumper sticker is a symptom of a larger change that has been under way for the last 50 years. For the 300 years before that, our culture was dominated by the science of Newton and Descartes and the deal that Descartes made with the Inquisition. Science took the material part and the church got the transcendent part. For scientists, that came to mean that material was real and anything that was not material was not real.

However, 100 years ago, the very material science of physics was shaken up with the arrival of quantum mechanics. Suddenly there was a non-material ground to our material reality. Worse than that, there were non-local connections. Scientific truths began to

Introduction

emerge that did not fit at all in the Newtonian material universe. Einstein called it "spooky action at a distance," and he did not like it.

Then in the wild '60s, popular culture in the West was introduced to spirituality from the East. About the same time, researchers (on the fringes of respectable science, of course) started looking at non-local connections between living things: Plants that knew their companions were being hurt, humans who can heal people by waving their hands over them.

And now, in the early years of the 21st century, the quantum revolution is creeping into the life sciences. The results are amazing. Back when physics was doing its quantum revolution, a lot of physicists were upset at the changes in how protons and electrons worked. But how many physicists do you know? Or how many protons do you see? The first quantum revolution produced a lot of popular science books on the topic, but there was not much change in how regular folks went about their days.

The results are going to be much more interesting as we mix quantum mechanics with the life sciences. The rules about what is possible in day-to-day life are changing. Things that were not allowed by science are becoming allowed: Things like healing at a distance, talking to a god, even the existence of god.

I trust that you, dear reader, recognize that I am getting a little ahead of mainstream, scientific reality with statements like that. I am talking about fringe science here. I think the fringe is on its way to becoming mainstream because of a basic rule of science: Any science, no matter how crazy, that accounts for the observations is automatically better than any science that does not.

The problem that we have with our mainstream beliefs, both in science and in the culture at large, is that we have been ignoring all the non-material stuff that Descartes ceded to the Catholic Church. People have always seen the future in dreams, healed with intent, talked to their mothers who died last year and talked to god. The list of behaviors that are not allowed under current science and medicine is very long, indeed.

When we apply the principles of quantum mechanics to the science of life, we get amazing, wonderful things. All of those behaviors become possible. The new science is doing a better job of accounting for the observations of the non-material parts of our lives.

I was in college at the University of Wisconsin in the '60s, so I was part of that influx of Eastern wisdom. I must confess that I was not directly in the influx since I was in engineering school. I wore my hair short. I went to all my classes and I only attended one protest. But those were formative years. I have spent all those years since then looking for spirit in science. Now, all these years later, I have found that with just a few, small steps past current scientific research, I can account for those very old, spiritual truths. All really is one. Ye really are gods. Spirit and life are quantum processes so the properties of the quantum field are suddenly relevant to how we live our lives.

Introduction

The journey of this book starts with the principles of quantum mechanics, proceeds through the new developments in the life sciences, to a place where we can validate the old spiritual truths.

A note about the image on the cover

I am going to refer many times to the influence that humans and gods exert in the universe as being *subtle*. The picture I chose for the cover is a very nice illustration of what I mean by "subtle influence." It is only a good illustration, however, if you understand how the image is formed. The picture on the cover and in Figure 1 is called a strange attractor[1]. The idea comes from chaos theory, the science that tries to describe apparently chaotic situations. It describes the outcomes of events over time.

Figure 1 A strange attractor

Each point on the curve (the shape is made up of many, individual points) is the state of a system, like molecules of air in a box with the bottom being heated, or the population of lemmings. If you take a series of measurements of the system, the individual results will appear to jump all over. They appear to vary randomly from one reading to the next. But if you plot the points, over time a pattern emerges, like the one on the cover. The curve is called an "attractor" because the apparently random variations in the state of the system are attracted to some shape. It is called "strange" because the shape is, as you can see, strange.

In that sense, the strange attractor is a very nice example of the kind of subtle influence that we will talk about a great deal in this book. So, if you are having a hard time with the idea of probabilistic influence that we will talk about later, just look at the cover and imagine changing the shape of the attractor.

About the structure of the book

The Parts begin where we are and progress to where we might be. Part 1 describes our current situation. It serves as a backdrop to the discussions that follow. Part 2 describes the science of applying quantum mechanics to life, specifically to thinking, feeling and health. Part 3 is concerned with the implications of the science in Part 2 for what is possible in human life. Part 4 takes a big step from talking about humans to talking about god, spirit and how it all works in the new science. Part 5 talks about what life might be like if people were to embrace the possibilities described in Part 4.

I should mention that Part 2 has a different style from the other parts. In Part 2 I'm in my "explaining science

Introduction

stuff" mode. I like to explain how science works. The other parts of the book are less technical, or more speculative and so they are a little more conversational in style.

I make a distinction between information and practice. Learning new information is very pleasant. I like it a lot. Sometimes, as is the case in this book, the information has implications for things we can do, that is, for practices. I understand that for many people taking up a new practice is a major change. My self-help background compelled me to include some information about practices to implement some of the possiblites we talk about in the book. My experience in scaring information seekers away by talking about practice caused me to put the short section on practice in the appendix.

The appendix also includes a section showing the equations for the Fourier transform and the quantum wave function. I appreciate that most people do not like to look at math, so I did not include the equations in the body of the book. The engineer in me compelled me to include the equations somewhere.

So, dear reader, read on and I hope you enjoy the journey.

---

[1] Images generated by Chaoscope, http://www.chaoscope.org

Part 1 Our Starting Place

Every journey begins where you are. That's not usually a problem for journeys in the physical world. The starting place can be a problem for journeys in the inner world. Many of us are not very good at knowing what really is going on inside.

The journey in this book will take us to a mostly old reality by a path through new science. It's a "mostly" old reality because the science adds some interesting new twists. We each start out, of course, from exactly where we are. I want to help us be clear about where that is, so I will describe the landscape of our current collective realities. I hope we all set off on the right foot.

Chapter 1
Building Reality

WE START OUR journey from a place where we know how the universe works. At least, we know how the mundane aspects of our little spot in the universe work. I know what to expect from my spouse, friends and co-workers. I know my knee hurts if I carry the laundry down stairs. I know my sister is having a hard time with her lupus. We know what we can and can't do. I know it would help if I did the knee exercises my doctor gave me. I know I can't help my sister. This set of "things I know" makes up my reality. It feels comfortable, solid, and, well, real.

If we are feeling adventurous, we can look a little beyond our own reality and find people that seem to live in a different reality. Some of those realities look better than ours, many look much worse. We don't usually inquire very deeply about why that is. I am going to begin our journey by looking at how we make our realities. It turns out that the comfortable, solid reality we wake up to is really quite flexible. This is useful information because what we believe about reality determines what we can do and be.

The ultimate subject of this book is changing the way we live. Change, of course, is never popular, so I would like to help make the change easier. It is one of my paradigms – much more on paradigms in a minute –

that if I understand something, then I can be comfortable using it. Or, to be a little less kind, if I think I understand something, I think I can control it. To understand how to change our current reality, we need to understand how we build our realities. Then we can talk about making changes.

We are going to talk about how the universe works. But that's not quite accurate. We are really going to talk about how we think the universe works, which is very likely different from how it really works. Regardless of what we think, the universe works quite nicely, thank you very much. It has churned along for a very long time. It managed to exist and develop before humans appeared and I am quite sure it will continue to exist after humans are gone. We humans try to figure out how things work and propose lots of stories, theories and paradigms. After making up the story, we usually proceed to assume that the story is reality and kill people who disagree with our story. But the story is never reality.

Reality as models

The universe works, but we don't really know how. The stories and theories we make up are terribly limited by our senses, by our instruments and by what we currently believe about how it works. The models of how the universe works have a long history of changing.

Making models

Every one has ideas about how the universe works at all levels and scales. We live in this universe, on this planet, in this culture, in this neighborhood, in these relationships. We do this kind of work; we have (or not) this kind of spiritual practice. We function in all these different contexts. The way we function is to have a model of how each context works so we know what to expect next and how to behave in each situation.

Human beings are every flexible. We can survive in all kinds of situations. The key to that flexibility is that we have flexible models. By flexible models, I mean that we humans can adopt all kinds of models. Once adopted, however, it becomes very hard to change models. It might be better to say that at formative times in our lives we can develop models that work for whatever situation we find ourselves in.

The model building process is one of trial and error. If the errors don't kill us, we take our interpretations and feelings about the current situation and add them to the model to use next time. So, our models are built from the "material" events in our experience, but those events are processed though the feelings, interpretations and meanings we associate with those events. It is really our feelings about an event that become our model for that situation. Feelings about a given event vary tremendously among individuals, so it is not surprising that our models can be very different.

But we humans are social creatures. We live in groups and need groups to be healthy. The models individuals hold about shared events need to be similar enough so that the group members can function without killing each other. You can probably think of several groups that don't get along. A look at the evening news will show you several places where groups are killing each other over differing views of how the world works. While they don't make the news, you don't have to look very far to find many groups that do get along reasonably well: communities, social and political organizations, even some schools and churches. So, many of the models that we use are shared with the people that we regularly interact with, in spite of the fact that we developed each of our models through our own personal interaction with the world.

Models as reality

The models we develop and use guide us in every aspect of our lives, from the most personal and unconscious to the most intellectual and analytic. This gets us into trouble on occasion because many of the models that guide our personal behavior were developed at a very early age, before we were five years old. In all cases the models are heavily influenced by feelings and emotions. These models are not based on any sort of "objective" analysis of external events and their outcomes. That's why there can be so many different models for the same physical situation. What is the "right" church to attend? What kinds of relationships are OK? What is good music?

In spite of this rather shaky foundation, almost all of us assume that our own models are real. More than that, people assume that their models <u>are</u> reality. This applies to Stone Age people and their belief that spirits occupy inanimate objects, to modern racists and their beliefs about the inferiority of other races and to doctors and their belief that homeopathy does not work.

Because people do take their models as reality, any threat or challenge to those models is taken as a threat to their reality. Heretics are treated badly in almost all domains, from religion to hard science. The combination of the arbitrary nature of the models and most people defending their models to the death make living in a time of changing models very interesting.

Changing models

The models are subjective, at best, and what is subjectively good at one time in one situation will not be subjectively good in another time and situation. This is true in all domains, including religion, social custom, business, politics, medicine and science. So, the models that people build change with time and conditions. Remember when the world was flat, or a woman's place was in the home?

It does happen that individuals can change their models on important issues, but it is much more common for people to adhere to the models they developed during their formative years and take the models with them to the grave. It is probably possible to find people for whom a flat earth and domestic women are still their reality.

When my father was in engineering school at the University of Wisconsin around 1940, his professor for general chemistry, who was certainly tenured, did not believe in atoms. When I asked my father about him, he said he remembered the professor saying, "...and they're talking about trying to split those things. Isn't that silly?" The professor retired, with encouragement, shortly after my father had his class.

The little town where I grew up had a large Catholic church. I remember when the rules were changed so that Catholics no longer had to eat fish on Friday. That seemed very strange to my youthful mind. Did God change his mind? How could something that was wrong last week, eating meat on Friday, now be OK?

Models in science: paradigms

Here in Western culture, we pride ourselves on being scientific. We insist we aren't superstitious and we reject ideas that aren't based on hard proof. What is considered part of reality and what is not real is determined by our mainstream science, at least for the mainstream of our culture. This adds considerable inertia to whatever the current model is.

Science likes to think of itself as objective, inclusive and exhaustive. That is, science thinks it looks at the world without prejudice or bias. It does not exclude anything, which means it looks at everything. If that is true, then saying that something is "not scientific" means that it is not real.

But the real reality is that science is not some abstract intelligence, it is people who are scientists. All people build their models of reality the same way, so science builds its theories and paradigms the same way mere lay people build their models.

Science is conducted in paradigms. Paradigm is a fancier word for model. The current paradigm in a branch of science defines what kinds of questions can be asked and what kinds of answers can be obtained, or, more specifically, what kinds of answers can be published. Phenomena that fall outside the current paradigm are not even considered. The result is that science is not objective, inclusive or exhaustive. Most scientists are as superstitious about their paradigms as any religious fanatic.

Science does have one advantage that lay people and religious fanatics do not have, and that is the quantification of the paradigms. Everything is measured and described in detail. The result is that scientific paradigms that restrict what can be found also provide a good mechanism for identifying anomalies, or things that do not fit in the current paradigm. These anomalies can be safely ignored early in the life of a paradigm, but they accumulate: unpublished, but not unknown. At some point, a few rogue scientists start looking at the anomalies and the current mainstream paradigm enters its declining years. Paradigms get old and die, just like the scientists who make them.

The historical trend has been that scientific paradigms change with increasing frequency. Ptolemy's flat earth model lasted 1300 years, until it was replaced by the physics of Newton and the cosmology of Copernicus and Kepler, which lasted 400 years, until it was replaced by quantum mechanics and relativity. The life sciences have managed to hold on to Newton's reality against the quantum revolution for another 100 years, which brings us up to the present. Our current era is a time of changing paradigms, both in science and in the culture at large.

Impact of change

The periods of stable reality between revolutions in the sciences are getting shorter. This means that the periods of stable reality for the whole culture are also getting shorter. The quantum revolution of the first part of the 20th century was revolutionary for physicists and astronomers, but did not have much impact on the day-to-day lives of the ordinary population. That is not entirely true. One of the practical results of the quantum revolution did have a big impact: nuclear weapons were developed. The Copernican revolution was more subtle. It changed how we view our place in the universe: We were no longer the center of the universe.

The paradigm change that is underway now is that quantum principles are invading the life sciences. This will change what we believe is possible in our day-to-day lives, so it could be a change bigger than moving from the center of the universe to the outer arm of a rather run-of-the-mill galaxy.

These changes are hard because reality is changing. People find it difficult to change their realities. Suddenly being able to eat meat on Fridays was probably not terribly upsetting to most Catholics. No longer being in the center of the universe was more upsetting, but the change was more a matter of faith and philosophy rather than how you lived your life. The quantum revolution in the life sciences that we are entering now will be a change in how we live our lives. Things that are widely believed to be impossible to do will become not only possible, but normal. This will be a big change. We can expect to have a very interesting time.

We have models for all levels of our lives: mundane to cosmic

I have described the models that science uses, the paradigms, as if there were just one, which is not the way it works, in science or anyplace else. Science, religion, business, culture and individuals all view their surroundings through the models and theories that are currently in vogue. And it takes many models, on many levels, to construct a reality.

People need a model about oral health: One person might brush and floss twice a day and another might brush once a week. They need a model about people who are different from them: Don't trust foreigners, or we are all brothers. They might have a model about outer space: The space program is a hoax, or space is a multidimensional gateway to other worlds. Most people have models for god and spirit: There's no such thing, or all is one.

When we talk about changing models, it makes a difference what level of model we are talking about changing. The smaller models are easier to change than the bigger models. Changing to brushing twice a day is a very different proposition from a change precipitated by losing a job. People have been known to die when they lose their jobs and cannot get a suitable new one.

The models do not have to be consistent or compatible

Humans have an amazing ability to keep their models for different domains of their lives completely separate. I suppose this contributes to our flexibility, but it makes for some interesting contradictions. There seems to be no need to have any sort of consistency between models. A person can behave one way in church on Sunday and quite another way on the job during the week. People can be very pro-life on the issue of abortion and very pro-death on the issue of war with our enemies. A company can treat its management employees with one set of principles and use quite another set of principles for dealing with hourly workers.

Along with keeping models separate, people are capable of ignoring huge amounts of evidence that does not support their current model. I mentioned the chemistry professor who did not believe in atoms on the eve of nuclear weapons. As I write this, there is an article in the New York Times about a small segment of the population that believes that the space program is a complete hoax ("Vocal Minority Insists It Was All Smoke and Mirrors", by John Schwartz, *New York Times*, July 13, 2009).

People do not seem to be very concerned about consistency or contrary evidence in their models. Why should they be concerned? We're talking about my reality here, and my reality just "is."

Models are changing

Where we are now is a time of changing models in several domains. The life sciences are changing as the quantum revolution catches up with them. Unlike the quantum revolution in physics of 100 years ago, the change in the life sciences has implications for what is possible in our normal, day-to-day lives. This implies a change in personal beliefs and values. Some of these changes may bring about changes in personal spirituality, which could spill over into changes in religious organizations.

Of course, just because science changes, does not mean that anyone has to notice, given the way humans build our views of reality. New models of reality may have to wait for one or two generations to die out before they can take root and spread to the mainstream of our culture.

Chapter 1 — Building Reality

What is "the truth?"

Reality, both scientific and personal, changes from time to time. If reality can change, then what's real? What is the TRUTH? We humans like stability and security. We like to know how things work, what the truth of the matter is, so we can be in control. Not knowing how things work means we are not in control. We don't know what will happen next.

So, what is the truth about how the universe works? My answer is that I don't know. I do believe that the universe works some way; I just don't know what it is. The universe is a pretty big place. Even at the height of our human powers, we are very small on the cosmic scale of things. We are like an ant on the launch pad trying to see the structure of the space shuttle.

But we humans persist in observing our surroundings and making up stories about how we think things work. The stories are not reality, but lots of people treat their collections of stories as reality.

What is the truth about our stories? I think Thomas Kuhn summed it up nicely in *The Structure of Scientific Revolutions*[1] when he said that the only indication of the quality of a theory is how much information you have to ignore, that is, how many observations do not fit in the model. This gives rise to the life cycle of stories, theories and paradigms. A new theory becomes accepted because the old one had too many anomalies piled up. The new theory enjoys a honeymoon of sorts and then people start noticing anomalies or inconsistencies. The anomalies accumulate while the theory is enjoying mainstream acceptance. Eventually, some, usually young, scientists start looking at the anomalies and formulating changes or replacements for the current theory.

Our realities are not the solid foundation of our existence that many assume them to be. Whatever we take to be reality will probably change at some point. The change is always hard, but it is also inevitable, given the subjective source of the theories. Of course, there are no other sources of theories.

---

[1] Kuhn, Thomas, *The Structure of Scientific Revolutions*, Second Edition, The University of Chicago Press, 1970.

Chapter 2
Our Current Models and Paradigms

WE ALL MAKE our realities from models that we construct from our experiences in the material world and our feelings about those experiences. Fortunately, we seem to be able to share enough of our models with the people we live with so that we can get along with most of them. Our journey to new ways of looking at the universe begins with our current models. They are our starting point.

Western culture is a big place. There are lots of people and many variations in models and beliefs about how things work and what they mean, but our culture is defined by a set of stereotypical models, or mainstream models. I would like to start our journey with a look at what those models are. In describing these models I don't want to imply that everyone in science or medicine, or the general public, holds these models exclusively. Models in all these domains are changing so there are people who definitely do not subscribe to the models that I will describe. It is also true that a single individual is fully capable of using conflicting, and even mutually exclusive models in different life situations. I will look at some of the alternate models toward the end of this chapter.

Let's begin with the models that drive most of the behavior in our culture.

Newton is alive and well

Our day-to-day reality is described pretty well by the physics established by Kepler, Galileo and Newton. These people laid the foundation for our modern science. We like to think of our modern science as being objective and not influenced by feelings and emotions, but our science was not founded in an atmosphere of objective calm.

Copernicus started the change by suggesting that the earth was not the center of the universe. He published his ideas in *On the Revolutions of the Celestial Spheres* in 1543, the same year he died. He waited until the end of his life to publish the work to avoid incurring the wrath of the Inquisition.

Kepler and Galileo, contemporaries who published their major works in the first years of the 17th century, are credited with giving us the idea that science should be based on observation rather than exclusively on reason.

Galileo did incur the wrath of the Inquisition and was accused of heresy for teaching that the earth moved while the sun was stationary. He was forced to recant his ideas and, according to legend, while he was recanting, he said under his breath, "And yet it moves." This story first appeared a hundred years after his death so it is not likely to be true, but it indicates the nature of the conflict between the emerging science of the time and the power of the Roman Catholic Church.

The science that shapes our current models of reality was very much shaped by the need to avoid treading on the territory claimed by the church.

Chapter 2   Our Current Models and Paradigms

Descartes formalized the arrangement between the church and science when he published his book on separation of mind/spirit from body in 1649. The deal was that science would concern itself with the material world and the church could have the transcendent, that is, the non-material world. This made it easier for Newton to publish his model of the clockwork universe, *Principia*, in 1687. The clockwork universe that we have from Newton is a model of reality that can be completely described by natural laws, expressed as mathematical equations. These equations can account for the movement of planets and comets, pendulums, apples falling from trees and cannonballs. Based on these successes, and the desire to avoid anything that sounded at all spiritual, it was assumed that everything in the material universe could be described by those equations. If there were material things that were not covered by the current equations, it was assumed that appropriate laws would be discovered shortly.

Newton's model, which is still very influential, tells us that living beings are included in the material universe. So living things can be completely described by the same kinds of laws that describe other material objects. Living things are currently understood as electrochemical processing systems. Biology today is really molecular biology. The paradigm is that life can be described as the interactions of molecules. Individual, living bodies, being material, are separated and isolated from one another. The only available connection between living bodies is via physical messages. Since living things are material, they can only be influenced by other material things. Material, as I am using it here, includes matter and the four forces (the strong and weak forces inside atoms, electromagnetism and gravity) that can interact with matter. Things that are not material, like spirit, thought, feeling, emotion and intention, cannot have any influence on material things, and so they are not real.

Consider these examples. When a doctor runs tests on a patient to find the cause of his or her problem and the tests show nothing, the patient might be told that the problems are, "all in your head."

"Mind over matter" is a well known phrase, but it describes an impossible thing, like a perpetual motion machine. Anyone claiming to be able to influence matter with his mind is assumed to be a fraud or crazy. I said that our bodies are material, so in Newton's paradigms, the mind cannot influence the body. This brings up a great example of our ability to ignore evidence that conflicts with our models: the placebo effect in drug tests. It has has been largely ignored by the medical mainstream. It is ignored in spite of the fact the 30% to 50% of people given the placebo get better.

If you are reading this book, you are probably not a hard-core material reductionist and you are probably saying, "But we all know that feelings and emotions have a huge effect on the material health of the body."

You are quite right, of course. What I mean is that our mainstream, science-based medical system does not recognize feelings and emotions as causes of material disease. The model is still very Newtonian: physical and chemical diagnosis followed by physical or chemical treatment. Even in quantum science, where the idea that the observer affects the outcomes is well established, it is still assumed that the outcomes of quantum level experiments are the objective results of the experimental conditions alone. In quantum science, "observation" usually means measurement by some instrument. The idea that the intentions and feelings of the experimenters can be a force in the experimental outcomes is definitely not a mainstream concept.

Living in Newton's world

The idea that matter and the forces that move it constitute all of reality is a good summary of the main paradigm for science. It is also the model for most people, although few people would describe it in exactly those terms. By main paradigm, I mean the model that guides peoples' behavior in work, professional and social situations. Even in religious contexts, non-material influences are relatively limited. Most people don't think of their reality in terms of models. It is reality. It is just the way things are. The way things are is that "mind over matter" does not work. What you think and feel inside does not affect anything outside yourself. People who hear voices and see visions are crazy. Feelings and emotions are not important. Reason and logic are important. We are isolated physical beings. We can only be with other people by sending messages or being in the same physical place with them. The only way we can know anything about other people is to talk to them. We can only see things with our eyes directly or in a picture.

Newton's spirit

Newton is credited with giving us the clockwork universe, where everything can be described by the equations of matter and motion. We think of Newton as the founder of mathematical science. But I think Newton got a bad rap. Newton thought of himself as an alchemist. Alchemy is an ancient practice that most people know as a means of trying to convert base metals into gold, or trying to find the philosopher's stone to give eternal life. At its core, however, alchemy was a spiritual practice, the polar opposite to the clockwork view of the universe.

Newton practiced alchemy for 30 years and wrote over a million words on the subject. I did not know that until just a few years ago. I had physics classes in high school and college and got two degrees in chemical engineering and never heard that Newton did alchemy. At the time of his death in 1727, alchemy was not held in high regard by the science establishment of the time. The Royal Society said that his writings on alchemy were, "not fit to print."

Newton's alchemy papers were rediscovered in the middle of the last century and the truth came out: Newton felt that there was more to the universe than could be conveyed by laws and equations, that the universe did not run like a big windup clock mechanism, ticking away with dreary predictability. He believed that there was something more to the universe.

I find that very encouraging. I am going to argue that there is much more to our reality than the clockwork model of the universe allows. It's nice to know that the nominal father of that model believed the same thing.

Descartes' deal is still in effect

Descartes lived after Kepler and Galileo and before Newton, 1596-1650. He worked actively in many fields, including mathematics and philosophy. He spent his whole working life being afraid of the Inquisition and not without reason. In 1633, the Inquisition condemned Galileo and Descartes prudently decided to abandon his plan to publish his book, *Treatise on the World*. In 1663, his works were put on the Index of Prohibited Books by the pope. Conditions had changed considerably by 1727 when Newton died. Newton's spiritual writings on alchemy were banned by the Royal society, and his scientific works were well accepted. Descartes is usually credited with helping to bring about that change.

Descartes argued in *Passions of the Soul* and *The Description of the Human Body* that the body is a machine and, so, follows the laws of physics, while the mind is not a machine and is not constrained by the laws of physics. This dual model of the human being provided a way to get the church and its Inquisition out of science. Science would stick to the machine body and the church could have the transcendent mind, the soul and the rest of the non-material stuff. The sciences, and especially medicine, have adhered scrupulously to that agreement.

The mind can clearly affect the body, by directing it to walk around and sing a song, but that was the only interaction between the mind and the material body allowed in Descartes' description. That means that things like knowing about someone's medical or psychological state without asking him or her is not allowed. Influencing someone's state using only your mind is also not allowed.

I have talked to a few physicians about energy medicine and the practitioners of energy medicine. Energy medicine is a form of mind over matter or action at a distance. It describes the practice of healing people using only intent. There is no physical interaction between healer and patient. Prayer is a well-known form of energy healing. Energy healing is clearly outside the current medical paradigms.

Now, doctors are usually very reasonable, logical people. I found it very interesting that when the topic of energy medicine came up, these reasonable doctors became quite unreasonable. Referring to the energy medicine practitioners, they said things like, "Put the quacks in stocks!" I have a pet theory about that, which is purely conjecture on my part. That kind of out-of-character reaction is inspired by the collective unconscious memory of the fear scientists felt about the Inquisition.

Whatever the mechanisms are that shape peoples' reactions, the mainstream paradigm in science and medicine is that the body is a machine and follows the laws of physics while the mind and spirit are not machines and do not follow the laws of physics. The body is a proper subject of science while the mind and spirit are not. Many members of the general public are also uncomfortable with any phenomena that seem to violate the Newtonian assumptions and Descartes' deal.

Patriarchy is still in charge

Patriarchy is defined as a culture where the male is the authority figure in the home and lineage is traced through male descendants. A large majority of the world's cultures are and have been patriarchies, at least as reported by patriarchal authors of history books. That somewhat benign definition of patriarchy does not mention the very widely known side effect that women are valued much less than men in patriarchal cultures, so the treatment of women ranges from just unfair to widespread abuse and violence. As a result, depression is twice as common in women as in men. Another side effect that is not so widely known is that patriarchy also assigns a very narrow role to men. Fitting into that narrow role causes considerable pain for men and the people around them. Since feelings and feeling disorders, like depression, are not allowed in men, men tend to act out their pain. This often takes the form of drug and alcohol addiction and the abuse of women and children.

This is not a book on social justice, and that is not why I brought up the patriarchy issue. I brought it up because the effect of patriarchy is to value the stereotypical male characteristics and to devalue and even demean the stereotypical female characteristics. The stereotypes are that men are rational, logical and take action based on conscious thought. Women, on the other hand, are dominated by irrational feelings and emotions and their actions are driven by those base sensations. This idea is deeply ingrained throughout our culture: Rational thought and logic are good and clearly superior to the untrustworthy and dangerous feelings and emotions.

The attributes that are valued in a patriarchy, like reason and logic, nicely support the reasonable mathematical model of reality in Newton's clockwork universe.

Hold this thought. It will be very important when we talk about the implications and applications of the model we will develop in the coming sections.

Religion

In a patriarchal culture, it is not too surprising to find that the main religions are patriarchal. Christianity is predominantly patriarchal. There are certainly minor exceptions to that rule, Quakers come to mind, but the mainstream denominations have varying degrees of male dominance. It is, of course, hard to emphasize logic in most religions, but the emphasis on male characteristics leads to an almost complete dependence on conscious language in Christianity. The bibles are read and prayers are said, all in conscious words. This way those messy feelings and emotions can be avoided.

There is another aspect of Christianity that is relevant to where we are going. If you look at spiritual practices around the world, both inside organized religions and outside, you will see a wide range of approaches. All these approaches can be arranged in a spectrum that is completely heaven-centered on one end and completely earth-centered on the other. In a heaven-centered spirituality, the emphasis is on heaven and the afterlife. The present physical life is not important at all, or it may simply be a means of preparing for the afterlife or earning a desirable place in the afterlife. Ascetics occupy the farthest reaches of the heaven-centered end of the spectrum. They ignore the body and its needs to focus on the spiritual and the afterlife. They may even abuse their bodies as a way of showing their disdain for the physical and earthly.

The other end of the spiritual practice spectrum is earth-centered spirituality. The focus is on the earth and nature as spiritual entities. More or less organized religions occupying the earth end of the spectrum in the United States are called pagan or Wiccan. Earth-centered spirituality was probably the first form of human spiritual practice.

Spiritual practices that include the body as part of the practice are closer to the earth end of the spectrum than to the heaven end. These are popular in the Far East and include the yogic spiritual practices described in the Vedas of India and the Taoist practices of China. The state of the body is an integral part of the state of the soul in these practices, which is why nurturing the body with healthy diet and movement like the yoga poses, tai chi and qigong are important parts of these practices.

Christianity, the mainstream religion in the United States, lives on the heaven-centered end of the spiritual practice spectrum. There were some ascetics who were influential in the early church, notably Augustine, who pushed Christianity away from any focus on the body or earth. Christianity grew out of the Jewish tradition and Christ, after all was a Jew. Most of the early practitioners and all of the leaders of what we now call Christianity were Jews, at least for the first 100 to 135 years after Christ's death. Judaism is another patriarchal religion. While Judaism is not a heaven-centered religion, it is certainly centered on language and learning.

The result of these influences is that Christianity is a cerebral religion. Conscious thought and words are used to contemplate heaven and what one has to do in this life to get to heaven. The body, its feelings, nature and earth do not figure prominently in the mainstream of Christian churches. We will see that this rational, cerebral, male model of reality is in for a major change in the quantum life paradigm.

Dualism

Like all other models of reality, the models of god have changed a lot over time. Before the rise of civilization, everything was inhabited by spirit: rocks, plants, animals, the stars, weather, volcanoes and people. When civilizations developed, the gods changed. They moved from being "in" everything to being local deities who lived on mountains or other sacred places. The Greek gods living on Mt. Olympus are an example. Then a Zoroastrian idea spread to the Middle East: there is only one god. The one god can't live anyplace local, so god moved to heaven, or at least someplace far away from everyday, material life.

Having god in heaven means that he/she may or may not be actively supporting any specific animal, person, group or region. So, what happened is that people began to distinguish between things that were "of god" and those that were not. Those that were not of god were considered bad and could be exploited and abused with impunity. They probably didn't feel any pain, anyway.

This situation was, and remains, common throughout the civilized world, from Old Testament stories about beheading believers in the wrong religion, through the Crusades, to the subjugation of indigenous people wherever they were encountered by advanced civilizations.

So we have one form of dualism: making a distinction between things that are of my god and things that are not.

And then Descartes came along and helped science get out from under the Inquisition by proposing another dualism: excusing the transcendent mind from the purview of science while making the body a physical machine. Now it's not just the "thems" of the world that are not of god, now my body is not of god, either.

Chapter 2   Our Current Models and Paradigms

In appearance, one person might look very much like another, but we keep track of whether one person is "us" and whether another person might be "them" so we can treat them differently. Repressing and controlling all those "thems" leads to enormous suffering. Maintaining these dualities and keeping track of all the "thems" out there takes a great deal of energy that could be much better spent.

These god-based dualisms are special cases of the more general problem of maintaining multiple reality models, which we take up next.

Multiple realities

We each live in a reality that is made out of a particular set of models of our own creation. Fortunately, many of the individual models are similar enough to allow groups of various sizes to get along, most of the time. When individuals do not share a common model about a situation, then they tend to not get along very well. This is because people believe that their "models" of reality are reality. And when that other jerk has some crazy view of the world, there's no way I can even talk to him. So, differing models of reality can cause problems between people, ranging from petty disagreement to violent hostilities.

I have described what I called mainstream models in the previous sections and I said that those models could account for most behaviors, which I believe is true. That does not imply that all those people whose main behaviors are directed by mainstream models have only one model of reality. The problem is that the mainstream models, given to us by Newton, Descartes and their contemporaries were developed in the crucible of the

Inquisition and the need to avoid its wrath. That means that large parts of the human experience, that is, all the non-material and spiritual stuff, were left out of the models. So, here we are in the 21st century, faithful disciples and intellectual descendants of Newton and Descartes, but we still experience the complete human and animal experience, which includes all that non-material stuff. In spite of over 300 years scientific materialism being the mainstream model of reality in our culture, there are still lots of churches and people going to them. There are tens of thousands of practitioners of non-material healing. All of us routinely use information that we did not receive through one of our five senses.

How do we cope? The answer is that we hold multiple models of reality and we use different models for different situations. Fortunately for our sanity, humans are quite capable of holding multiple and even conflicting models. The hard-core Newtonian doctor can be quite comfortable going to church on Sundays and participating in prayers for the recovery of his fellow parishioners who are in hospitals many miles away and feel that the prayer helped those people. Another way to describe the situation is to say that we have dual, or more than dual, models of reality: one set for the material side and another set for the non-material.

It takes a certain amount of energy to keep the conflicting models separated. If the situations, like work and church do not overlap, the process is quite manageable. It gets interesting when the situations begin to overlap, for example, when the Newtonian doctor is asked to include prayer as an important part of his treatment of a patient, or when the Newtonian husband is confronted with his wife's intuitive insight about a career choice or an investment strategy.

All of the models are quite arbitrary and they are all subject to change over time, so the idea of any sort of "absolute truth" is not at all realistic.

Life here on earth is made very interesting by the fact that almost all of us, from professors of physics to members of the flat earth society, assume that all of our models are true and absolute representations of reality. People can ignore vast amounts of information and experience that fall outside their model for that situation.

Keeping these models straight and properly isolated, deciding which model is appropriate for each situation takes a lot of energy. Then there's all the effort in selecting what to ignore, and resolving the internal conflicts between incompatible models and the external conflicts with other individuals who hold conflicting models.

Our lives are filled with, mostly unconscious, separation, isolation and conflict. All this conflict takes its toll on our personal mental, physical and spiritual health. It also gets expressed in the abundant conflict we see in the world around us, which extracts a high price in human suffering.

But nothing ever stays the same for very long. In fact, our culture is entering a phase of changing models. You can see the initial skirmishes between Newton's paradigms and the more spirit-centered models in medicine every day.

There's more than the mainstream

Our mainstream models of reality are decidedly Newtonian and Cartesian (after Descartes, as in Cartesian coordinates on graph paper). I am pretty sure that most people have built at least part of their reality from Newton's ideas, but that leaves room for many other realities.

There is a small but growing part of the population that seems to be less constrained by the Newtonian view of the world. These people regularly practice energy healing, medical intuition, remote viewing, psychic readings and the rest of the paranormal activities that are not allowed under the mainstream models. Then there is the much larger portion of the world's population that believes in a life after death of the physical body, souls, angels, demons and god, in many forms. This group even includes people who are members of the scientific mainstream paradigm when they aren't in church.

Our starting point is a place where our cultural norms disallow a big part everyone's experience. We cope with mechanisms that range from ignoring the offending events to holding conflicting models of reality, to living completely outside the mainstream culture.

Change is in the air

Our starting point also includes ongoing change in science and the culture at large. My personal model of how cultural reality works in the West is that science trickles down to popular culture. Science, in my reality, also works on a trickle down process: Change trickles down from the fringe to the mainstream. The quantum and relativity revolutions of 100 years ago were revolutionary in a few fields like nuclear physics, astronomy and cosmology. "Revolutionary" is not consistent with "trickle down," but the quantum ideas did not have much impact on other branches of science, so I can still claim that new science generally trickles down from the fringe to the mainstream. Up to now, the quantum revolution in physics has not had much effect on how people live their lives, except for the nuclear weapons.

The life sciences, biology and medicine, have largely resisted any inroads from quantum effects. We will see later that small things like electrons, protons and even atoms behave very differently when they are in the quantum state compared to when they "collapse" into the normal material state. When I say that the life sciences have resisted inroads from quantum effects, I mean they have resisted using atoms in that quantum state to describe life processes. The reason given has been that living systems are too "warm, wet and noisy" to allow any quantum behavior, that is, any particles in the quantum state immediately interact with other particles and collapse into the normal, material state. But the trickle down process is working. Quantum mechanics has been successful in describing many details of molecular interactions in living systems, hence the appearance of quantum biology.

Quantum biology is concerned mostly with molecular-level interactions, just like the rest of biology. I want to talk about human-level functions, like perception, feeling, memory and intention. The application of quantum processes to these large-scale behaviors in living systems is definitely a fringe proposition. There are people doing work in these areas and they are reporting their results, but the work is far away from mainstream acceptance. The hint that this fringe work may be an accurate description of how things "really" work is that the results can account for many phenomena that mainstream science, lacking a Newtonian explanation, chooses to ignore.

Those changes in science are trickling into popular culture. As evidence I offer that changing market share of conventional and alternative medicine. For some time now there have been more visits to alternative practitioners than to general practice physicians. Enter "quantum" or "spirituality" in the Amazon search bar and you will get long lists of books.

This book is an attempt to speed up the trickle down process that moves new information from the fringes of science to how all of us lay people live our lives. Most of us live our lives according to Newton's models of reality. By doing so, we are ignoring many capabilities that lie outside Newton's model (at least the model that history chose to preserve).

Letting quantum mechanics into the life sciences, and having the results trickle down to the general culture can accomplish several worthwhile things. We can ease the burden of maintaining models of reality that require us to ignore a big part of the human experience. It will be OK to be open to all experience.

We, as a culture, will gain capabilities that we have denied for 350 years. They include the ability to heal others and ourselves. We can be healthier. If we recognize the power of self-healing, then perhaps we can move past the machine model of health and trying to fix spiritual and psychological problems with chemicals.

If the impact of quantum mechanics turns out to be anything like what I will describe in these pages, then we will find that spirit and god have returned to everything. Everything and everybody is of god. Everything is us. This could save a great deal of wear and tear on humans and our environment.

What is really real?

Where we are now is that the scientific mainstream, particularly in the life sciences is still adhering to Descartes' deal with the Church, while a small, but significant segment of the science community pursues research into phenomena that are clearly outside the mainstream paradigm. This fringe research is motivated by the large numbers of people who routinely practice things that are not allowed in Newton's model of life. Those practices include medical intuitives, energy healers, psychics, people who pray, fortunetellers, remote viewers, shamans and monks of all sorts. Of course, not all practitioners of those arts are equally effective, and some are really frauds. But then, you can say the same thing for practitioners of the mainstream healing practices, too. All sorts of ordinary people have premonitions about the future, dreams that give them meaningful and useful information, see deceased relatives, have unity experiences, feel spiritual presence, get useful information from talking to beings that no one else can see and on and on.

There is a growing number of card-carrying scientists who are willing to explore these out-of-the-mainstream-paradigm phenomena. They do their work, for the most part, very carefully because they know their mainstream colleagues will challenge it. Their work is published in obscure journals because it is outside the paradigms of the big-name journals. Sometimes it is only published in a book. There are lots of popular science books that summarize this fringe work.

Here are some books and their topics. This is not an exhaustive list. These are books that I have or am familiar with, in no particular order. There are lots more. Most give references to the research, if you want to dig deeper.

Christopher Thompkins and Peter Bird, *The Secret Life of Plants*[1], on Cleve Backster's work with non-local communication between plants and humans

Lynne McTaggart, *The Field*[2] and *The Intention Experiment*[3], overviews of research in all these topics

Larry Dossey, *Healing Words*[4] on the effectiveness of prayer and *The Power of Premonitions*[5] on our ability to know the future.

Dean Radin, *Entangled Minds*[6] reporting his research in non-local phenomena

Rupert Sheldrake, *The Presence of the Past*[7], on morphic fields and non-local influences

William Tiller, *Conscious Acts of Creation*[8] reporting his research in the effect of intention

Michael Talbot, *The Holographic Universe*[9] on research and historical evidence for paranormal phenomena

Charles Tart, *The End of Materialism*[10] on scientific evidence for paranormal phenomena

Chapter 2 — Our Current Models and Paradigms

Tom Harpur, *Uncommon Touch*[11] on evidence and experience with energy healing

Mona Lisa Schulz, *Awakening Intuition*[12] on medical intuitives and intuitive knowing

The position I take in this book is that all that experience and all that research is sufficient to demonstrate the reality of these paranormal phenomena. I am not going to try to convince you that this stuff really happens. If you need convincing, the books I listed and many others do an excellent job of demonstrating the reality of these things. I want to talk about how and why they work.

If we accept that the paranormal is part of reality, then, from a purely scientific point of view, any model that accounts for those phenomena is automatically better than any model that does not. There are not many theories on how these things work, so most anything I suggest that is reasonably coherent is a defensible scientific model.

After I build some models, we can look at what they mean for how we live our lives.

So, let's begin at the top, with the science. I'll let nature take its course for the trickle down process.

---

[1] Tompkins, Peter and Bird, Christopher, *The Secret Life of Plants*, Harper Paperbacks, 1989

[2] McTaggart, Lynn *The Field*, HarperCollins, New York, 2002

[3] McTaggart, Lynn, *The Intention Experiment*, Free Press, New York, 2007

[4] Dossey, Larry, *Healing Words*, HarperCollins Publishers, New York, 1993

[5] Dossey, Larry, *The Power of Premonitions*, Dutton, 2009

[6] Radin, Dean, *Entangled Minds*, Paraview Pocket Books, New York, 2006

[7] Sheldrake, Rupert, *The Presence of the Past*, Vintage Books, New York, 1988

[8] Tiller, William, *Conscious Acts of Creation*, Pavior Publishing, Walnut Creek, CA 2001

[9] Talbot, Michael, *The Holographic Universe*, HarperPerenial, New York, 1991

[10] Tart, Charles, *The End of Materialism*, New Harbinger Publications and Noetic Boks, 2009

[11] Harpur, Tom, T*he Uncommon Touch*, McClelland & Stewart Inc., Toronto, 1994

[12] Schulz, Mona Lisa, *Awakening Intuition*, Three Rivers Press, New York, 1998

Chapter 3
Science and Spirit

THERE IS ONE aspect of our current condition that deserves special attention: the science versus religion issue. We saw in the last chapter that these two domains were defined as mutually exclusive paradigms. That partitioning of reality reinforced the spirit versus body division that the Christian church had been enforcing since Augustine's time, the end of the 4th century. This division has been a big issue in my life and, as I am sure you have noticed by now, is a major motivation for this book. I would like to look more closely at the issue.

The long separation

So here we are, 350 years after Descartes made machines of our bodies and Newton made a machine out of the universe. For all these years, science has scrupulously avoided the church's transcendent turf, that is, the non-material phenomena. Matters of spirit and things that happen across distance with no physical communication are not proper subjects for science.

I will confess to being at least optimistic, if not naive, about lots of things. Here's an example: It seems to me if the purpose of science is to examine and explain reality, then anything and everything in reality should be a proper subject for scientific study. The non-material phenomena have always been a large part of the human

Life and Spirit in the Quantum Field

experience, and they remain so in spite of 350 years of trickle down from material science. Therefore, science should be happy to objectively examine healing and spiritual issues along with dodos and Higgs bosons (hypothetical particles that give other particles mass). But, then, as Snoopy famously observed, "We aren't playing should'ves."

We have a mainstream scientific model that says there is no spirit in reality. We also have lots of historical experience that tells us that models in science, and everything else, change from time to time. The normal mode of change is that observations outside the mainstream start to appear at the fringes. They grow and seep toward the center.

I heard a rather cynical description of the change process that went like this: Mainstream scientists first ignore the new results and models, then deny them, then persecute them, then allow that they might be interesting, then agree that they are real, but that it is old news. Whatever the process is, spirit and the non-material are growing in from the fringes toward the center of modern, materialist science.

Getting reacquainted

I think the first crack in Descartes' deal came with the quantum revolution at the beginning of the last century. Quantum mechanics and relativity did not follow the usual "creep in from the fringe" model. They went from conception to nominal mainstream status (at least in nuclear physics and cosmology) between 1900 when Max Planck proposed quantizing energy (that's where the quantum in quantum mechanics came from) to 1926 when the Copenhagen interpretation of quantum

| Chapter 3 | Science and Spirit |

mechanics was issued. Quantum and quantizing energy refers to energy coming in discrete chunks rather than a continuous stream, like ice cubes instead of the stream from a hose. Einstein's relativity papers, published in 1905 and 1916, contributed to the revision of Newton's science.

It was quickly observed that scientists talking about quantum weirdness sounded a great deal like Taoist priests (We will talk about that in the next section). Neils Bohr, a very prominent figure in the development of quantum mechanics, put the Taoist yin and yang symbol in his family crest. The non-material seed was planted.

In the 1970s, two books came out that explicitly examined the connection between spirit and science: *The Tao of Physics*[1] by Fritjof Capra and *The Dancing Wu Li Masters*[2] by Gary Zukav. The similarities between the words Taoists use to describe their universe and the words quantum physicists use to describe their universe are striking, but up until recently there was little work that could be used to describe how the phenomena of spirit worked in scientific terms. The problem has been that the domain of quantum mechanics – photons, electrons, quarks and such – is (or seems to be) far removed from the domain of human inner experience.

Science and Inner Experience

In Western science, human inner experience is not a consideration, but there are cultures in the world where the human inner experience is part of scientific inquiry. The Buddhist, Taoist and old Vedic cultures in the East come to mind. The science, of course, is quite a bit different from science here in the West, but it satisfies the requirements of a science. They do experiments, observe the results, propose models or theories and test the theories. The difference between this Eastern science of the spiritual and our science in the West is that in the East, the subjects of the experiments are always the inner states of human beings. The instruments used to measure those states are always words in the form of questions and answers and descriptions. In other words, the human perception of the human inner state is taken to be a valid measurement of that state. In the West, we like to think that the human inner state is not relevant to any right and proper scientific undertaking.

Quantum mechanics and life

As long as quantum mechanics was only concerned with little things, subatomic particles and such, the old models used outside of quantum mechanics were safe.

> A note on definitions: The terms beginning with, quantum, and ending with mechanics, physics, or theory, are all interchangeable. I think quantum mechanics is the official name for the branch of physics that uses quantum principles to describe the behavior of matter and energy. I like the term, quantum mechanics, so that is what I will use in this book.

But like everything else, that is changing. Back in the '70s, scientists started looking at non-local phenomena in living systems, things like the interaction between humans and plants (that was Cleve Baxter in the '60s, actually) and energy healing or healing at a distance. Biologists started finding quantum effects in the chemistry of living things. In the '80s, British biologist, Rupert Sheldrake, started publishing seriously fringy theories about a morphic field that guides the form and function of all things across time and space.

Biology, these days, is really molecular biology: the study of life at the molecular level. Newton (in his clockwork model) and Descartes would approve. Most of the applications of quantum mechanics to living systems are at the molecular level. This is interesting, but it does not help us to get where we want to go, which is describing the highest levels of human functioning. Only a few people in science, like Baxter and Sheldrake, have attempted to talk about human level functions in any sort of non-local terms.

New science

I believe that it is possible for a form of Western science to embrace the non-material and non-local (and non-temporal) aspects of the human experience. I think it needs to if we are to eliminate the dualism in our lives that takes up so much of our energy and contributes so much to the conflict and suffering in the world today.

Of course, the science that does embrace the non-material will probably not look very much like our current science. The change from a materialist science to something bigger will make the quantum revolution in physics look like a Sunday picnic in the park.

Today, there are almost enough pieces of science available to describe the non-material, non-local and non-temporal aspects of human experience. I should not limit the discussion to humans. All living things, indeed, all material reality, have those same non-material, non-local and non-temporal connections.

Science that includes spirit

Jumping ahead, I can say that it is the holographic nature of the quantum field that makes those connections possible. After we say that, we have to do some work on some deeply held assumptions in order to move forward. One assumption is that we, you and me, are separate, material objects that walk, talk, think and feel on our own, isolated from other material objects. Another assumption is that our material brains are the source of our thoughts, feelings and decisions. These assumptions are the heart of the Newtonian view of reality, but they do not provide a complete description of the universe. The quantum mechanics view of reality, which has a spectacular record of successfully accounting for the way little, material things work, can be applied to human-level function. It tells us that our non-material parts, thoughts, feelings and such, originate in the quantum field, the same place that electrons and protons originate. We will talk a great deal more about the quantum field in the next chapter. Those thoughts and feelings are projected into material reality through our senses.

After we get past Newton's view of the world, it is only a small step to explain how the universe thinks and feels on its own. And there, dear reader, we have god.

Ye are gods

We can explain how god works by applying quantum principles to how life works. That is very interesting, but I was amazed by what came next. After I worked though how we exert influence through the quantum field, it appeared that we mere humans exert exactly the same kind of influence as the entities we call gods. The scale is different, but the influence is the same. We actively influence how things happen all over the world. And it's not that we do this once in a while, or that only special people do it. Every living being does it all the time, continuously. We'll see that what you are feeling is your continuous contribution to the ongoing creation of the world. That includes the feelings you have, but are not aware of, because they are buried in your shadow. Scary, huh?

Being a god is a big responsibility. Wouldn't the world be a better place if more people were aware of that responsibility and took it seriously? It is not just me, or even you and me, who get to be gods. Everyone gets to be god. Everyone contributes to the ongoing creation of our world.

If this all sounds radical or extremely unlikely, I should point out that this is not an original idea. The idea that humans are like god is very old. It appears in the Bible in Psalms 82. "Know ye not that ye are gods?" It appears outside the Christian tradition, as well. We will come back to this idea later.

Our current science and our current religions hold seeds that can grow into a single, coherent view of our universe. It will, of course, take a lot of care and feeding for those seeds to grow up through the many layers of paradigm and belief carefully designed to keep science and religion (spirit, really) separate. But then, I'm ever the optimist.

That will be quite a trip considering our starting point here in a materialist world where people feel free to exploit other people who are less worthy. Even the longest journey begins with the first step. Our first step will be to look at some thoroughly secular, somewhat arcane science. Let's begin.

How far have we come?

Our first steps on the journey have shown us that reality is not as solid as we might hope. It is made out of models that we make up and learn in response to the things that happen to us. We have lots of models for all the different situations in our lives. Those models are often inconsistent and contradictory.

The big institutions in our culture, religion and science, also build their realities from models. Some of those models are quite old and outdated, but institutions, like the people that make them go, are very good at ignoring information that conflicts with a current paradigm.

We will see why we depend on these subjective models when we get to the chapter on perception.

We are good at maintaining conflicting models inside ourselves, but we don't do well with other peoples' conflicting models. Science and religion are a good example of that conflict. I jumped ahead of the story when I explained how science might be compatible with religion if science would recognize the human inner experience and if religions stopped trying to be different from every other religion, and making everyone else wrong. We'll talk about a science of spirit in Chapter 14.

---

[1] Capra, Fritjof, *The Tao of Physics*, Shambala Publications, 1976

[2] Zukav, Gary, *The Dancing Wu Li Masters*, William Morrow, 1979

Part 2 Applying Quantum Mechanics to Life

Our journey starts exactly where we are. So, here we are in a culture with a strange mix of conflicting paradigms. We are a science-based culture. The nuclear physics part of our science is based on quantum mechanics that has all these non-material, non-local and non-temporal ideas. The life sciences are still based in Newton's materialist paradigms, but quantum principles are creeping in from the edges. None of the sciences are comfortable with anything spiritual, religious, paranormal or anything related to any sort of mind over matter.

In spite of the opinions of science, religions and spirituality flourish in our culture. Practitioners of the paranormal arts abound. Three quarters of the population believes in the paranormal. Medicine, which tries very hard to be scrupulously scientific, is losing market share to complementary, alternative and paranormal practitioners. We live in interesting times.

Almost everyone maintains more than a few paradigms that are not consistent with each other. Our journey here is to try to reduce paradigm conflict. To do that, I want to take some of the principles of quantum mechanics, mix in some new models of how our minds and our physiology work, and arrive at a paradigm of how life works that makes sense of the current conflicts. This is our path in the coming chapters.

Chapter 4
Everywhere in the Hologram

WE LIVE IN a universe that seems to be two different places. One place is full of isolated material things wandering around and occasionally bumping into each other, like billiard balls bouncing around on a pool table. The other place is also filled with material things, but they are connected across the time and space that separate the material bodies. Knowing and influencing take place without any bumping required. There are even non-material things that have intelligence, but don't have a material body. The healing, spiritual place seems to be very different from the day-to-day, material world that we all occupy. In our normal world, things are only here *or* there, never here *and* there. We never bump into ourselves. My mistakes are etched forever in the past. Our lives seem to be over in a twinkling and then we are gone. The only way I can find out what my brother is doing is to call him on the phone. The only way I can help a friend is go visit. Reality ends with what I sense, with my bare senses or my instruments. It all feels so reassuringly secure and concrete.

The non-material world that some people seem to occupy is very different. Things can be everywhere at once, the past and future are malleable, we can exist across time and space (at least some part of us). People can know things across time and distance without phones or wires or any physical aid. People can be healed across time and space, the healer does not have to be there. Reality is much larger than the five senses can perceive, and that larger reality is perceivable. You can even talk to it. But it all feels very unsure, indefinite and fuzzy.

The underlying science for these effects is the hologram. The math that underlies the hologram is the Fourier transform. It plays a key role in the mathematical description of the quantum field, but that is not the only reason I will bring it up here. The Fourier transform can also be used to describe the form of our thoughts and feelings, so it will play a very important role in connecting our human, inner state to the field.

The Quantum Field

These days almost everyone has heard the terms, quantum field and quantum mechanics. The Q word appears in a great many book titles. The subject itself is much worse than rocket science. It is a very complex domain both in its concepts and in the mathematics that describe it. Fortunately for me, and the readers of this book, I don't need to do real quantum mechanics. There are just a couple of aspects of quantum mechanics that we need in order to account for our nonmaterial phenomena. Those two aspects are the holographic nature of the quantum field and the probabilistic behavior of quantum level events. So I'll talk about the hologram in this chapter.

Chapter 4 — Everywhere in the Hologram

The nature of holograms

I said we were interested in the holographic nature of the quantum field. That statement is only interesting if we know what holographic implies, so we begin with holograms and why they are the way they are.

Holograms are photographs taken without a lens. In a projection hologram a 3D image is produced when you shine a light through the film. There are also reflection holograms that produce different images when viewed from different angles. You probably have some of those on your driver's license and credit cards.

The piece that interests us about holograms is that information in a hologram is distributed, that is, it is everywhere. We can account for that property with the math that is used to describe holograms. That bit of math is the Fourier transform. Let's look at what it is and how it works. I know that bringing up math is not a good idea in books for the general public, but this particular piece of math plays a big role in several places on our journey, so I'm willing to take the risk and talk about it. I am not being completely insensitive to my math-phobic readers. There are no equations in any of the text. For the curious, I have put some of the equations in the appendix. You can only see them if you open up the appendix.

Fourier transform

The Fourier transform is a bit of math named after Joseph Fourier (1768-1830), who was a mathematician and was appointed governor of lower Egypt in Napoleon's Egyptian campaign. Fourier showed that functions can be described as the sum of a bunch of sine waves.

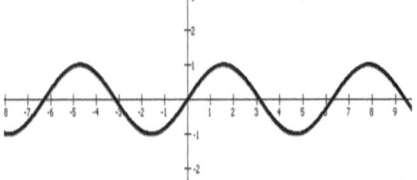

Figure 1 A sine wave

I assume everyone is familiar with sine waves. Figure 1.

Sines are big in trigonometry, as you may recall. They also describe many things associated with rotating

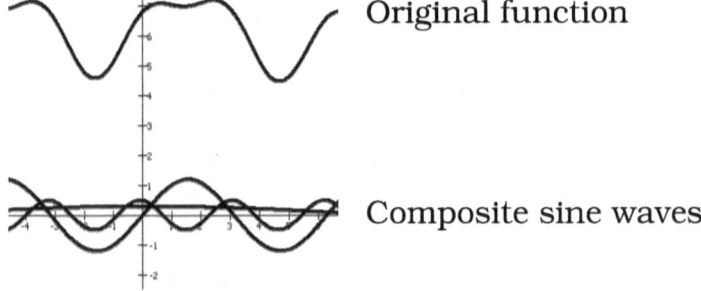

Figure 2 Describing a function as sine waves

movement. The sine and cosine are also functions and they are periodic: they cycle through the same values over and over.

Fourier showed that any function can be described as a sum of sine waves. If you pick the sine waves correctly, you can lay them out and add them up. The sum will be the original function. Figure 2.

Chapter 4 — Everywhere in the Hologram

That seems like a nice little math trick, something you can use to amuse your friends on a rainy afternoon. Well, it turns out to be more than a nice math trick, a lot more. At the practical level the Fourier transform is used in signal and image processing, electronic circuit design, cell phone transmission, oil exploration, quantum mechanics and more. At a more esoteric level, we will use the Fourier transform to talk about how feeling and soul and god work.

The reason the Fourier transform is so important is that it does a good job of modeling, or describing, many of the properties that we are interested in, like how we can know things across any distance, or how the soul can exist across all time. To see how that works we need to

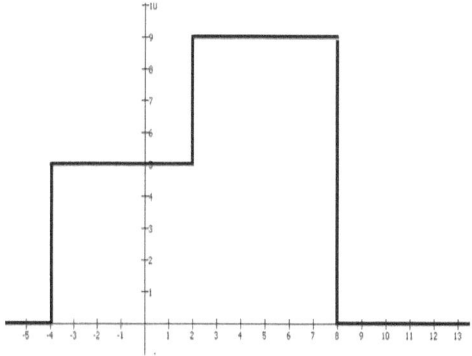

Figure 3. A simple function

understand the Fourier transform as another function and not just as a collection of sine waves. The transform function defines those sine waves, but I can draw it as another line on a graph.

Here's a simple function, Figure 3.

And here is its Fourier transform, Figure 4.

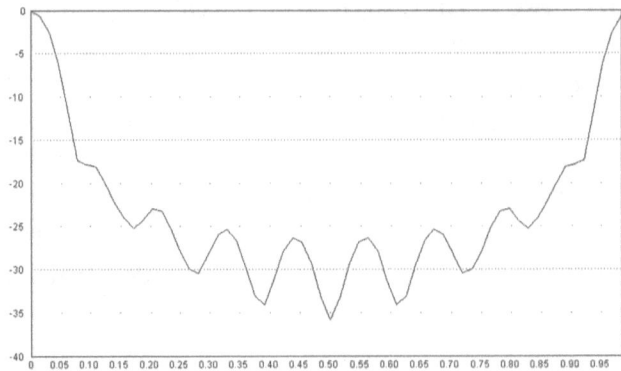

Figure 4 A Fourier transform

The Fourier transform takes information in the real domain and distributes it across the domain of the transform, usually called the frequency domain. It does this in two ways.

First, in order to calculate the Fourier transform of a function, you have to do an integral from minus infinity to plus infinity on the real function. Or, in regular language, I have to do a calculation at every point in my real function to get the Fourier transform function. This means that there is no one-to-one correspondence between a point in the original function and any one point in the Fourier transform. The second way the Fourier transform distributes information is that the coordinates of the axes are inverted. If I have a function of time, when I take the Fourier transform it is a function of one over time, or frequency. For this reason the Fourier transform side is often called the frequency domain.

The equation for calculating a Fourier transform involves an integral, exponentials and imaginary numbers. For those of you who don't mind looking at the math I've shown the definition of a Fourier transform in the appendix. For those of you who do mind looking at math, you can leave the equation in the appendix without looking at it.

Information in my real function that is close to the origin is far from the origin in the Fourier transform. Information that is far from the origin in my original function is close to the origin in the Fourier transform.

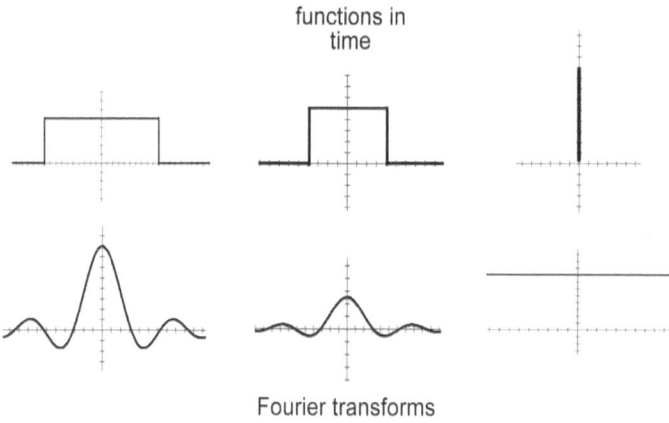

Fourier transforms

Figure 5. Inverse relationship between time function and the Fourier transform.

As a result there is a inverse relationship between a function and its Fourier transform. In the picture, Figure 5, my time functions are on the top, the Fourier transform functions are on the bottom.

As you move across the page the time function becomes narrower in time. The corresponding Fourier transforms become wider. There is an inverse relationship between a function and its Fourier transform.

The point of all this is to demonstrate that the information in my starting function is distributed, or smeared around, in the Fourier transform.

Light and Fourier Transforms

Fourier transforms are not just a bit of abstract math. It turns out that they describe light.

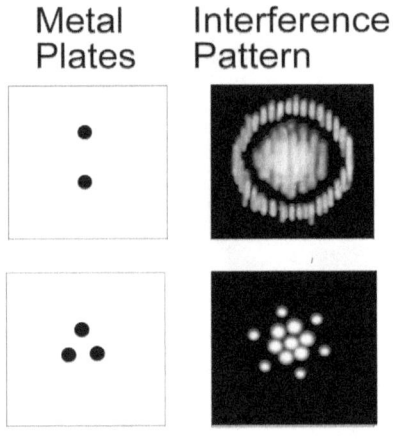

Figure 6 Light and Fourier transforms.

Figure 6 shows some white boxes with some black shapes in the middle. Think of the white boxes as little metal plates with the black shapes being holes punched in the plates. If I shine a laser on one side of the plate and put a piece of film on the other side of plate, I will record the image produced by the light coming through the holes in the plate. The black squares to the right of each white square show the images recorded on the film for each plate. These images are classic interference patterns and are well known phenomena in all kinds of waves. If I were to calculate the Fourier transform of the patterns in the metal plates and plot the function,

Chapter 4 — Everywhere in the Hologram

actually the square of the function, I would get exactly the same image as I recorded on the film. The Fourier transform does an excellent job of describing the behavior of light.

That leads to a very simple and physical description of what the Fourier transform is. If you look at something without a lens you are seeing the Fourier transform, or the frequency domain. If you look at something through a lens you are seeing the inverse Fourier transform, or spatial domain. A lens does an inverse Fourier transform. Bear in mind that if you are looking with your eyes, there is a lens. The hologram we will see in our Jack and Jill example is what you see when you look at something without a lens. Look ahead to Figure 10.

Light and lenses provide a nice example of non-local information. Consider the lenses in your eye. If your eyes are working correctly you see a smooth and complete image of the scene in front of you. If you take two steps to the right you still see the same scene, but point of view has changed a little bit. No matter where you move, you see the complete image. The amount of light you take in through your pupils is a very small fraction of total light available. But no matter where you sample the light with your eyes you see the full image. All the information about the scene is everywhere in the light. When light is reflected off all the surfaces in a scene, it is taking a Fourier transform. When the light passes through the lenses in our eyes that is taking the inverse Fourier transform.

So Fourier transforms are a bit of math that take information from our regular space and transform it into a frequency domain. In doing so it spreads out the information from our regular space all over the frequency domain. Far from being a bit of abstract math, the Fourier transform is a precise description of how light and lenses perform. This is the first step on our path to understanding the holographic nature of the quantum field. Next I would like to look at holograms. It turns out that Fourier transforms can be used to describe holograms.

Holograms

I assume that most people are familiar with holograms. They are the three-dimensional images that float in space in front of the film. If you recall the first Star Wars movie when Princess Leah sent a warning to Luke, it was in the form of a moving hologram. He turned on the little box and an image of Leah appeared in the air above the box, slightly translucent, but talking and moving. That was a three-dimensional, animated hologram.

Holograms were invented by Dennis Gabor in 1948. He calculated the behavior of coherent light using Fourier transforms and found that he could form a three-dimensional image on film and then use coherent light to project that image back into space. At that time there were no good sources of coherent light. He had to wait until the early '60s when lasers were invented to make a hologram of a reasonably-sized object. He won the Nobel Prize for his work.

Holograms will be useful on our journey to the quantum field because they provide a couple of properties that we need. More importantly, we can say that the quantum field is a hologram. So let's consider holograms.

Chapter 4   Everywhere in the Hologram

Making a hologram

Let's consider a piece of physical reality, say Jack and Jill on a hill, Figure 7.

Figure 7 Jack and Jill on a hill

Jack has fallen down and is at the bottom of the hill. Jill is still up on the top of the hill. In physical reality we say that they are separated by time and by distance. We can measure the distance between them. We say they are separated by time because any communication between them requires time. If Jack sets off a firecracker Jill will see the smoke before she hears the bang because light covers the distance between them much faster than sound.

Now, let's take a hologram of the scene. You can't take holograms of outside landscapes yet so imagine that we make a very detailed ceramic model of our scene. Now we can put the ceramic model in a box and take a hologram of it, Figure 8.

We put the model in a box and we put a piece of film over in one corner. From the other side of the box we introduce a laser beam. The beam is split in two. Half of the beam is directed via mirrors straight to the film. The

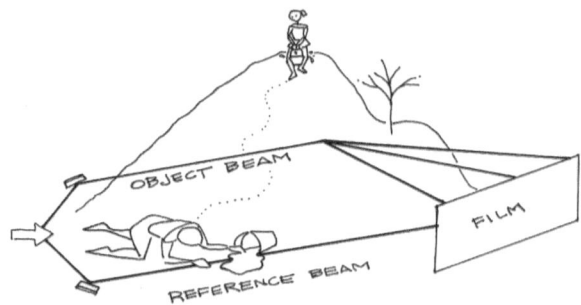

Figure 8. Taking a hologram

other half is directed to our ceramic mountain scene where it bounces off the model and then hits the film. The part that goes straight to the film is called the reference beam and the part that bounces off our figure is called the object beam.

The waves in the object beam get displaced by bouncing off the various surfaces of our little mountain scene. Because the laser light is coherent, that is, all the waves in the beam start out the same, when the two beams come back together in the film they form an interference pattern. When two peaks come together, they add and form a bright spot. When a peak and a trough come together, they cancel each other out. The information about where the peaks and troughs of wave are relative to each other is called phase information, Figure 9. It turns out that phase information is important to holograms.

Chapter 4   Everywhere in the Hologram

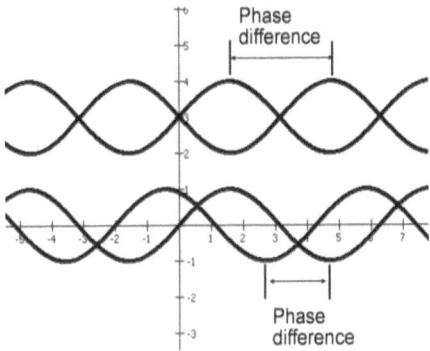

Figure 9 Phase of sine waves

The purpose of the reference beam is to convert the phase information into intensity information. Photographic film only records brightness, or intensity of light. The silver oxide in film, or the charge coupled devices in digital cameras, only respond to the amount of light that hits each spot. Film is like a bucket collecting rain water. The level in the bucket tells you nothing about when the water arrived in the bucket. In order to reproduce the image from the holograph in the film we need the phase information. The interference pattern produced by the reference beam captures the phase information in the object beam and converts it into intensity information so the film can record it.

Properties of a hologram

The film records that interference pattern. The laser light exposes the film and we can develop film and look at it. It looks like this, Figure 10.

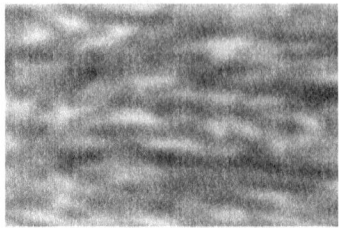

Figure 10 A hologram

It looks like nothing at all. There is no image in the film. But if I take a laser and shine it through the film at the same angle as the reference beam, I will get my mountain scene projected in space in front of the film, Figure 11.

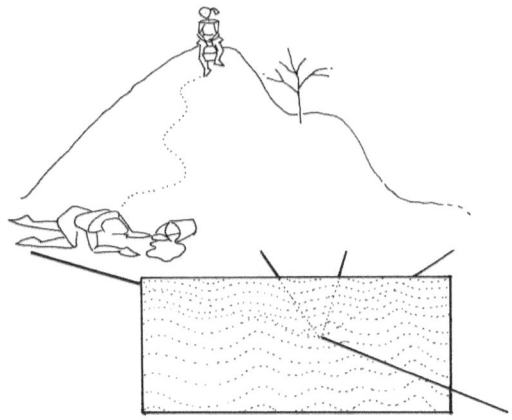

Figure 11 Projecting a hologram

It is a three-dimensional image in that if I move my head up and down the point of view on the image changes. I can't see the back of the mountain. That would take more exposures and more pieces of film. It's actually like looking at a scene outdoors through a window. I can move to different places and look out the window at

Chapter 4   Everywhere in the Hologram

different angles and see the scene outside from slightly different angles.

There are several properties of holograms that are relevant to our journey toward the quantum field. The first is that a hologram is a Fourier transform of the geometry of the scene. In 1948 Gabor calculated how they would work by using Fourier transforms. So we can say that a hologram is a Fourier transform of a piece of physical reality.

The next useful property is that holograms can carry a tremendous amount of information. After I do the exposure of my mountain scene I could remove the mountain scene and replace it with a teddy bear. I could change the angle of the reference beam a little bit and then expose the teddy bear with a laser light. I could then remove the teddy bear and replace it with a model car, move the reference beam a little more and expose the model car. In conventional film this would result in a multiple exposure where the three images are superimposed on one another. When I develop my triple exposed hologram and shine a laser beam at the angle of the first reference beam, I get my mountain scene. If I move the laser beam to shine at the angle of the second reference beam, I get the image of my teddy bear. If I move the laser beam to shine at the angle of the third reference beam, I get an image of my model car. The three images can be carried in the same hologram and not interfere with one another. Holograms are capable of carrying a tremendous amount of information. We will use this property a couple of times in coming chapters.

Non-locality

The last property, and the one that is the most important for us, is that the information in a hologram is non-local. That is, it is everywhere in the hologram. In real life Jack and Jill are isolated from one another by time and distance. But in the hologram the information about Jack and Jill and their separation is everywhere. I can show that this is the case with the classic hologram non-locality demonstration. Take the hologram, shine a laser through it and you get a complete image of our mountain scene. I tear the hologram in half and shine a laser through one of the halves. What do I get? The whole image. I take one of my halves and tear it in half again leaving me with a quarter of the original. I shine a laser through my quarter piece and I get? The whole image. I take my quarter and tear it in half two more times and shine my laser through the smallest remaining piece. What do I get? Still, the whole image.

While Jack and Jill may be separated in real life the information about their positions, the distance between them and about who they are and what they're wearing is everywhere in the hologram. If you have one piece of the hologram you have everything. You don't have to go anywhere or look anywhere else in the hologram because all the information is wherever you are. This idea that information is nonlocal and is distributed everywhere in the hologram is central to the holographic nature of the quantum field.

In the interest of accuracy I should mention that the images we get from smaller and smaller fragments of the original hologram are not exactly the same as the image we get from the whole hologram. It works like viewing a scene through a window. The image we see with the

Chapter 4   Everywhere in the Hologram

complete hologram is like the view we see looking through the whole window. We have quite a bit of room to move our point of view and see different angles on the scene outside. Tearing the hologram in half is like covering half the window: We can still see the same scene, but we don't have as much room to move our point of view. By the time we get down to a very small piece of the hologram, it is like looking out at the scene through a tiny peephole in an otherwise covered window. We can still see the same scene, but we only have a single point of view.

To put this another way, a hologram works just like the light that we sample with our eyes. If we can move around, we can change our point of view of what we are looking at. If we cannot move around, we only have one view of the scene in front of us.

It is also not entirely true that the information is everywhere. There is a size below which the information is not available. That size is related to the wavelength of light and it is very small.

A hologram, then, is a different kind of domain from our normal domain of space and time. It stores information about our normal space domain and we can reproduce our normal domain as a projection through the hologram. Information in the hologram domain is nonlocal. It is distributed everywhere in the hologram. And finally, holograms can carry a great deal of information.

With these properties in mind we are ready to take a very large step up to the quantum field.

Chapter 5
The Quantum Field

QUANTUM PHYSICS WAS a huge paradigm change for physicists at the beginning of the 20th century. The feeling at the end of the 19th century was that physics pretty well had things wrapped up. One prominent physicist at the time said that the physics of the future would take place in the fifth decimal place. He meant that all of the good work had been done and there was nothing left to do but refine the existing models. Besides refining the models, there were just a few little anomalies that needed to be tidied up and fit into the existing theories. Well, that's not the way it worked out.

Anomalies in physics

There were three small anomalies. One was called the ultraviolet catastrophe. This concerned the behavior of things when you heat them, things like ovens and pieces of steel. Metalworkers for as long as people have been working metal know that when you heat a piece of metal it first turns a dull red, and then becomes brighter red, and then becomes orange red, and finally moves to white heat. The whole practice of tempering metal is based on the fact that the temperature is always indicated by the color of the glowing metal. The same color is the same temperature, no matter what kind of metal is being worked. The problem was that the science at the end of the 19th century said that when you heat metal or an oven it should begin to emit ultraviolet light (which is not visible) and become infinitely bright as the temperature is increased. This prediction clearly did not match the observation.

Another anomaly was the way the photoelectric effect worked. It was known that if you shine a light on certain kinds of metal you get a current to flow out of the metal, that is, electrons. This is how solar cells work in your watch or calculator. The problem was what happened when you changed the intensity of the light. In 1902, it was observed that the energy of the electrons increased with the frequency, or color, of the light. The science of the time predicted that the energy should increase with the intensity, or brightness, of the light.

The third problem concerned atoms. Around 1900, there was still some debate about the existence of atoms. But among those who thought that they existed, atoms were understood to consist of a positively charged nucleus with negatively charged electrons orbiting about the nucleus. The problem with this view was that Maxwell's equations, which did a wonderful job describing electricity and magnetism, said a moving charge produces a magnetic field. Producing the field takes energy, so the electrons should eventually lose energy and fall into the nucleus. In other words, atoms could not exist. Since matter was made of atoms, then clearly atoms did exist. And in fact they seem to exist for a very long time.

Beginning in 1900, people began resolving these anomalies. In 1900, Max Planck proposed Planck's law that said the energy of a photon is a constant times the frequency of the light. He was not directly interested in any of the anomalies at the time. In 1905, Einstein had a very big year. Among other things he proposed that by using Planck's law he could resolve the ultraviolet catastrophe by suggesting that the energy coming off of a hot object can only be released in small bundles, called quanta. When he did this, he found that his calculations matched the observations of the Bronze Age metal workers.

Einstein's other 1905 papers proposed special relativity, demonstrated the existence of atoms and resolved the problem with the photoelectric effect. The paper on the photoelectric effect showed that by quantizing the energy of light, his calculations matched the observed behavior of photoelectric current. He won the Nobel Prize for that paper.

In 1912, Niels Bohr looked at the problem with atoms and found that he could explain the stability of atoms if electrons were limited to specific energies and specific orbits. He said that electrons could only have discrete, specific amounts of energy. Those discrete amounts of energy are called quanta. In other words, he quantized electrons. Over the next 14 years, a great deal of work was done and the result was that the Copenhagen interpretation of quantum physics was announced in 1926. It was a very strange science and there were lots of ways to interpret the math. They had a meeting and agreed on how it should be interpreted.

Quantum level behavior

Quantum mechanics explains the behavior of matter and energy that make up most of the material world that the great majority of us are accustomed to living in. The quantum mechanics equations tell us that things with lots of mass behave just like Newton said they would. That's encouraging because Newton did a pretty good job of describing the way our material world works.

But at small scales, quantum mechanics predicted some very strange behavior. Here, small scale means things the size of a single atom, protons, neutrons, electrons and photons. At this scale, things are downright weird compared to the world we are accustomed to living in. There is a body of experiments called "quantum weirdness," because the observed results are so entirely weird. More on that in the next chapter.

I should point out that the quantum equations are far and away the best description we've ever had of how matter and energy work. The quantum equations agree with experimental results to as many decimal places as we can measure, which today is quite a few. And while some of the predictions made by quantum mechanics are truly weird, they have all been verified through scientifically accepted methods.

The weirdness involves things that appear to be separated by time and space actually being connected and influencing one another instantaneously. It also involves transferring information and influence backwards and forwards in time. These kinds of phenomena are simply not allowed by Newton's physics and they are well beyond the normal experience of most people living in our Western culture. But these phenomena are not so foreign to people in other cultures, particularly those in the Taoist, Buddhist, yogic and shamanic traditions.

Chapter 5 — The Quantum Field

The quantum field: an addition to Newton's reality

The quantum field is a hologram of our material world. The equations that Bohr, Heisenberg and the rest developed to describe electrons and protons look like Fourier transforms. If we believe that the equations describe a physical reality then there is an additional component to Newton's reality. This new component consists of non-material waves. It is called the quantum field.

The field will provide the non-temporal and non-local medium for our nonmaterial phenomena. I'm also going to suggest the field provides all of the non-material intelligence and creativity that we see in our world.

The quantum field domain is profoundly different from the material world that we walk around in everyday. It consists entirely of waves. All matter and energy exists in the quantum field. If we want to calculate what's going to happen in the material world, at least at the level of atoms and photons, we need to do the calculation in this quantum field and then see how that gets projected to the material world.

God does play dice with the universe

The problem with making the projection from the quantum field into the material world is that the quantum equations do not tell us the exact outcome of any event. They give us the probability distribution. In order to see if the equations are "right," we have to run lots of experiments. We will get lots of different results (Figure 1).

Figure 1 Results under a probability curve

If those results are distributed according to the probability distributions given by the quantum equations, then the equations were correct. This probability thing was very upsetting to many physicists.

Physicists are accustomed to calculating exactly how an experiment will come out, as in, "The ball will land exactly there." They do the experiment and measure where the ball landed. If it was "exactly there," the theory was correct. Equations that predict exactly how an experiment will turn out are called deterministic. Most scientists and engineers deal with deterministic equations. Newton's laws are all deterministic, for example. The idea that the fundamental equations of molecular physics in quantum mechanics were probabilistic was very upsetting to them. Einstein put it nicely when he said, "God doesn't play dice with the universe." Actually, Einstein raised several objections to quantum physics and he was shown to be wrong on all counts.

Chapter 5 — The Quantum Field

Probabilistic outcomes

This notion of probabilistic outcomes will be very important to our later considerations of influencing human behavior, so I would like to look a little closer at what it means. Consider this experimental setup, in Figure 2.

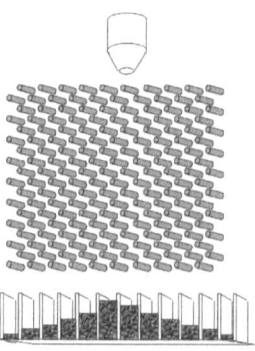

Figure 2 An experiment with probabilistic outcomes

We have a board with rows of little pegs in it, like a piece of Masonite pegboard with a dowel stuck in every hole. At the top of the board is a container full of little metal balls. The container has a little door on the bottom that I can open and let the balls run out. At the bottom of the board is a row of little boxes. The boxes have glass fronts so that we can see what's in each box. If I open the door on the container and let the balls out, the balls will bounce off the pegs as they fall down toward the little boxes. Since the balls were dumped in the center of the board, most of the balls will find their way more or less down to the center at the outlet and will collect in the center boxes. Some balls will get bounced around a little more and will spread out toward the edges and those will be collected in the boxes at either end. On average, most of the balls will be in the center boxes and relatively few will be in the boxes at either end. The way each individual ball bounces around in its journey down through the pegs is a very random process. That means that if we could mark one of the little balls with a red dot we could not predict which box it will end up in after it falls down through the pegs. We say that the outcome of any specific event, of a specific ball falling down through the pegs, cannot be predicted. We can however predict the probability curve. That is, we can predict how many of the balls will fall into any one box relative to the other boxes. In this case, the probability curve is a classic bell curve that everyone remembers from the grade curve in school.

When I say that the outcomes of quantum events are probabilistic, I mean they are like a little ball ending up in a specific box. I cannot predict or calculate where any specific ball will end up. I can only say that after you run many balls down through the pegboard, they will fill the boxes according to the bell curve. I can calculate the shape of the curve quite precisely and when all the balls have been collected they will be distributed in the boxes precisely as the bell curve predicted.

So, we say that the outcome of any one experiment, dropping a single ball down through the pegs, is probabilistic. If the outcomes were deterministic, then we could calculate exactly where any specific ball would end up. A key feature of the quantum description of reality is that the outcomes are only given as probability distributions, or probability curves.

Chapter 5   The Quantum Field

Nature of the quantum field

I have said that the quantum field is waves, just like the waves in a hologram. I then spent the last few pages saying that the quantum equations describe probability waves for the outcomes of events. It is a crucial step on our journey to recognize that the field is both: a hologram of our material world and the waves that determine what happens in our material world. Indeed, if we are willing to look at Schrodinger's famous equation for the electron, which I put in the appendix for those of you who would like to look, it looks a great deal like the formula for the Fourier transform. In fact, the quantum field equations can be derived using Fourier transforms. They were not developed that way originally, but I think it nicely demonstrates the similarity between the quantum field and a hologram[1].

The waves in the quantum field hologram are exactly the probability curves for the outcomes of events.

Getting to everywhen

Comparing the quantum field to a hologram makes the important point that the quantum field is the frequency domain of our material, time domain world, but a photographic hologram is a very limited example of the quantum frequency domain. The quantum field differs in some very significant ways from the holograms that we make on film. The film hologram is a static recording of geometric information in our material world. The quantum field, on the other hand, is the living, dynamic source of our material world. The field is not a recording of reality the way a hologram records our mountain scene. It's the other way around. The material world that we walk around in is a projection of movement and information that lives in the quantum state.

The lens analogy might be helpful here. I said earlier that we could understand a Fourier transform, or the inverse Fourier transform more accurately, as the action of a lens focusing an image. If we see movement in our focused image, that movement is not originating in the image. The movement is originating in the source of the light that the lens is focusing. The same is true of our material reality. The movement and growth and change we see everyday in our lives does not originate in the material world. It originates in the quantum frequency domain. What we see and experience every day is a projection of information in the quantum field through the lens of our perception.

The quantum field is a hologram of everything. In Newton's view of the world, space was considered separate from, and not related to, time, which is what ticks by on a clock. Einstein demonstrated that the two are connected. He showed that the real coordinates of our universe are space and time. Actually, space-time was first proposed by Hermann Minkowski. Minkowski was one of Einstein's teachers in Zurich in the late 1890s. So, if the quantum field is a hologram of our material reality, then it is a hologram of both space and time. It is not anything like the static image we have in photographic holograms of physical objects. The physicist David Bohm described the quantum field as containing holo-movement. A photographic hologram is a Fourier transform of static, spatial information only. The quantum field is a Fourier transform of space-time. It includes time, so things are "moving" in the quantum field. In our hologram of the mountain scene, all of the information about Jack and Jill was everywhere in the hologram. In the quantum field hologram of our space-time, all the information is four-dimensional: It is everywhere and everywhen.

---

[1] Tiller, William, Dibble, Walter, and Kohane, Michael, *Conscious Acts of Creation*, Pavior Publishing, Walnut Creek, CA 2001, page 240

Chapter 6
Reality and Quantum Weirdness

TO GIVE YOU a feel of how the quantum field works, I would like to describe some of the actual experiments done to demonstrate how things with little or no mass behave in the quantum field. I want to be a little careful here because these experiments do not account for how living things use the field. There have been a number of attempts in popular books to do just that. The argument goes that because a pair of electrons or photons can be entangled across time and space, that accounts for the entanglement of human thought and feeling across time and space. I don't think that explanation is sufficient to account for the entanglement of living things. I hope to give a much better explanation of how living things are connected later.

The experiments

For now, I would like to describe how the quantum field is different than our normal everyday reality. The easiest way to do that is to describe the collection of experiments that have actually been done called quantum weirdness.

The two slit experiment

Thomas Young did this experiment in 1801 to help resolve the debate about whether light was a particle or a wave. He took a thin metal plate and put two parallel slits in it. He shined a light at one side of the plate and put a screen a short distance behind the plate. If light is composed of particles, you would expect that the pattern on the film would be two little slits. If light is a wave, you would expect to get an interference pattern: a bright spot in the middle with fringes of decreasing brightness extending to either side. The experiment showed the interference pattern. This helped swing scientific opinion toward the wave model of light. Prior to that, Newton's particle model was more popular. The wave model held sway until Einstein proposed quantizing light, and then quantum mechanics suggested that light is both a wave and a particle.

The double slit experiment has become a classic demonstration of quantum effects. The light source can be reduced until just a single photon at a time is being released (Figure 1).

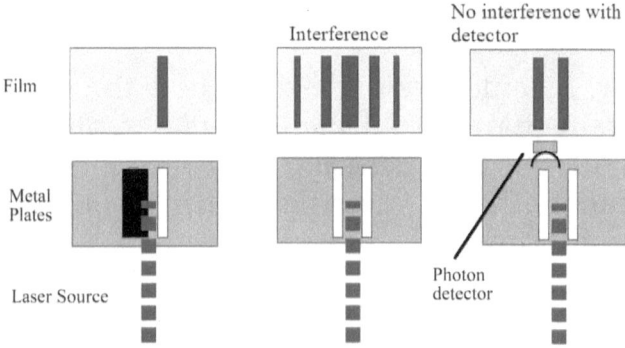

Figure 1 The double slit experiment

Chapter 6 — Reality and Quantum Weirdness

If we think about this experiment in particle mode, then a single photon has to go through one slit or the other. If it does that, there should be no interference. We can record the individual photons on a piece of film placed behind the plate. Each photon will expose one tiny piece of the film. If we keep shooting photons, eventually we will build up a complete image on the film. If there is no interference, we should see two little slits on the film, as shown in the right-hand image in Figure 1. But, what actually happens after shooting individual photons for a long time is that we get the interference pattern as shown in the center picture in the figure. Each individual photon seems to be going through both slits and interfering with itself.

You might say to yourself, "Well the explanation is simple: Light is really a wave and a single wave can easily interfere with itself."

But this is quantum mechanics and things don't usually work the way you think they do. It is possible to add a photon detector to one of the slits, the image on the right. This is a device that can detect the passage of a photon. With this device we can tell whether a photon goes through the left slit or not. Keep in mind that this detector does not interfere with the light in any way. What happens is shown in the right-hand picture of the figure. The interference effect is destroyed. The individual photons now behave as particles that pass through one slit or the other. They produce simple images of the slits on the film, and there is no interference. By simply knowing where a photon is, we change its behavior from a wave to a particle. If the detector is removed, the interference pattern returns.

Life and Spirit in the Quantum Field

Note that even when the single photon is behaving as a wave in going through the slits, it always behaves as a particle when it hits the film. Only a single point on the film is exposed.

Photons have no mass, so it is not too hard to think of them as behaving like waves. Quantum mechanics tells us that everything has a wave existence in the quantum field. The double slit experiment has been used to demonstrate the truth of that statement. Objects with mass can be shot at the plate and their landing recorded on film behind the plate. Electrons don't have very much mass and they produce an interference pattern even when a single electron at a time is shot at the plate. The same results have been obtained with protons, which are 1800 times more massive than an electron.

Whole atoms consisting of many protons, neutrons and electrons have been shown to interfere with themselves in this experiment. The largest particles to show this wave particle duality are buckyballs[1]. Buckyballs are molecules of carbon made up of 60 carbon atoms. They are arranged in a sphere and the placement of each atom looks like a geodesic dome, which was invented by Buckminster Fuller. In his honor, the molecules are named buckminsterfullerines, called buckyballs for short. A single buckyball is pretty small, but with a molecular weight of 720, it is huge compared to a single proton, which has a molecular weight of one, or an electron, which has a molecular weight of $1/1836$, or a rest mass of 9.1×10^{-28} gram.

The double slit experiment nicely demonstrates the wave particle duality of things with no mass, photons and even things with mass, from electrons up to buckyballs. The prediction from quantum mechanics is that

Chapter 6   Reality and Quantum Weirdness

everything is a wave in the quantum field. The double slit experiment also demonstrates one of the most upsetting aspects of life in the quantum universe, at least for Newtonian people: observation affects the outcome of experiments. And it is not even an actual observation. It is simply the ability, or potential, of making an observation. A prime assumption of all the sciences has been that we can observe the outcomes of experiments without influencing those outcomes. Quantum mechanics tells us that at the quantum level, observation is influence. This property will figure prominently in our discussion of influence and creativity.

The two path experiments

Now consider another bit of quantum weirdness in Figure 2.

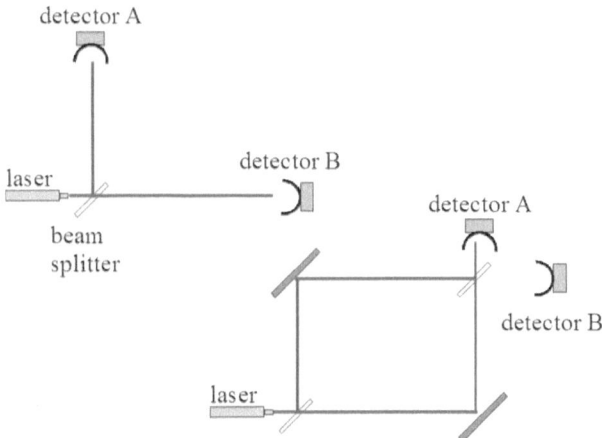

Figure 2 A simple two path experiment

In this experiment, laser light is directed at a half-silvered mirror. Half of the light that hits the mirror is passed through and the other half is reflected. If the mirror is set at 45 to the beam, the incoming beam will be split into two beams diverging at 90 to each other. If we put photon detectors at the end of each beam, each detector will receive half of the light, as shown in the upper left of the figure. If we reduce our laser beam down to a sequence of single photons, only one detector at a time will record a photon. That is, the mirror either reflects the photon or it passes straight through. This is all normal enough.

If we expand our experimental setup by replacing the detectors with full mirrors set at 45 , our two beams will be directed back together. If we put a half silvered mirror at the intersection and put our detectors in the paths of the two beams that can emerge from the half silvered mirror, we will have the setup shown in the lower right of the figure. When we shine laser light into our new setup, only one of the detectors will receive any photons. The reason for this is that the laser light is coherent and when the two beams come back together they interfere with one another. The result is that one of the beams is completely canceled out and photons only go to one of the detectors. Let's say that the detector at A is the one that gets the photons.

Based on our experience with the double slit experiment, what would you predict will happen if we turn the beam down to a single photon at a time? If you said we still get the interference and only one detector receives photons, you would be correct. A single photon introduced to our experimental setup appears to go around both paths and to interfere with itself when it gets back to the second half-silvered mirror. Photons will still be recorded only on detector A.

Chapter 6   Reality and Quantum Weirdness

We can change our experiment a little bit to produce an interesting result, Figure 3.

We take a thin metal plate with a hole shaped like a star cut out of the middle. We put that in the top beam just before the second half silvered mirror. We add a piece of film in the beam that usually does not get any photons. Now we introduce a single photon at a time into our apparatus and run that for a while.

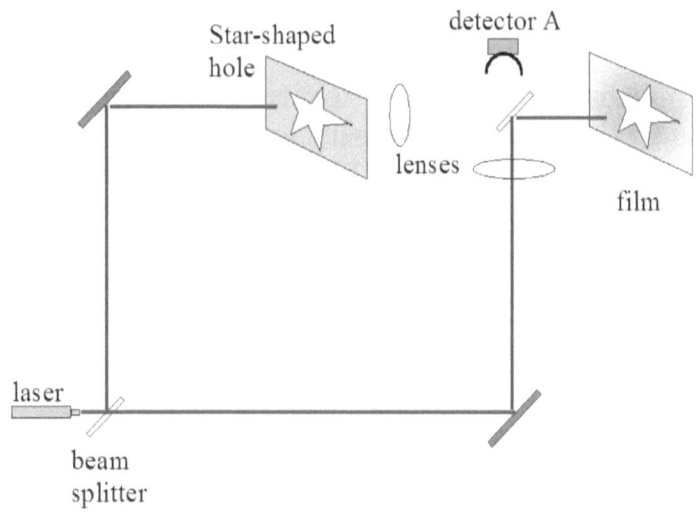

Figure 3 Two path experiment with a star

Remember that in our first set up, no photons will go to where the film is because they always interfered with one another in that direction. This time after we run photons through the apparatus for a while and develop the film, what do we see? What we see is a negative image of our metal plate with the star i.e., the film is exposed outside of the star, and unexposed inside the star. What happened?

Each single photon takes both paths through the apparatus. If the photon happens to go through the hole in the plate, then it interferes with itself and only goes to the detector side. Photons that go through the star-shaped hole will never hit the film. If the photon happens to hit the metal plate, it is blocked. That means that the part of the photon that took the lower path will not meet any interference and can be reflected to the film. But because that photon is really the same as the photon that hit the plate, it ends up on the film in the same place the upper photon would have been if it had not hit the plate. What that means is the only photons that can end up on the film are those that were outside the hole in the plate. The result is that the star itself is unexposed on the film and the space around the star is exposed.

I'll say that again. The single photon goes both ways around the square. Without the plate, each photon would interfere with itself so that we would only detect photons at detector A. When we add the metal plate and the film, sometimes the part of the photon traveling in the upper path will hit the plate and sometimes it will go through the star-shaped hole. When it goes through the hole, it will interfere with itself and we will only see it in detector A. When the upper part of the photon hits the plate, there will be nothing to interfere with the part of the photon coming up the lower path, so that photon will not meet any interference and it can be reflected to the film. But that part of the photon is "the same" photon that hit the plate in the other path and so it is in exactly the same place as the blocked part. The result is that photons coming around the lower path and never encountering the metal plate with its star-shaped hole, can produce a negative image of the star.

Chapter 6   Reality and Quantum Weirdness

The lesson of this experiment is that the two "parts" of the photon traveling on separate paths in the experiment are really the same photon. They are not separate "parts". There is only one photon. It is everywhere it can possibly be in the experiment while it is in the quantum state.

These kinds of results are very difficult to understand if we think of two separate particles going around each half of our setup. They appear to be separated by distance so how can one know what the other is doing? In our conventional world, the answer is that they can't and the results we see are impossible. But these are real, observed, scientific results and they are consistent with the predictions of quantum mechanics. These experiments are nicely described in *The Road to Reality* by Roger Penrose[2]. What quantum mechanics is telling us is that when things with low mass are in the quantum state, or in the quantum field, they are everywhere that they can possibly be, all at once. There is no separation of time or distance between things entangled in the quantum field, even if it looks like they are separated in our view from the material world. We will use this property of the quantum field to account for connection between humans and other living things.

These two experiments have demonstrated the connection of quantum objects across space. But, the quantum field distributes time as well as space. This next experiment demonstrates that time effect.

The Einstein-Podolsky-Rosen experiment

Werner Heisenberg was a major contributor to the development of quantum mechanics. One of his contributions was the well-known Heisenberg uncertainty principle. He said that certain pairs of properties of particles were coupled and they were coupled in such a way that if you measured one property with great accuracy, you could not measure the other property at all. The math behind this is that the two properties are Fourier transforms of one another -- our friend, the Fourier transform, again. As our knowledge of one property gets more specific, the value of the other property gets less specific.

The coupled properties that are easiest to understand are position and momentum or position and velocity. In our everyday world, position and velocity seem to be pretty much independent of one another. "I'm going exactly 65 mph and I am exactly at mile marker 231 on the freeway." And, indeed, this is not an effect we can see in our everyday world. But in the quantum world of very small particles and very precise measurements, if we know an electron's position very exactly, then we cannot determine its velocity or momentum at all. There are other pairs of properties of electrons and photons that are coupled in this way.

The idea that measuring one property on a particle affects the measurement of another property on the same particle was very upsetting to most physicists. Then it got worse. When Schrodinger's wave equation for electrons came out, the one that looks like a Fourier transform, it implied that two particles could be "entangled" across time and space. Entangled means that properties of the particles are coupled. If particles A and B are entangled and you measure a property on particle A, then the value of the same, or a coupled property, on particle B will be determined. According to quantum mechanics particle B would "know and respond" to the measurement on particle A instantaneously across any distance.

Einstein did not like this at all. It appeared to contradict one of the basic tenets of relativity: There can be no communication faster than the speed of light. If particle B responds to a measurement on particle A instantaneously, regardless of the physical distance between the two particles then, Einstein reasoned, information must be transferred from A to B faster than the speed of light. He believed this because his view of reality was the local material view, like our mountain scene in our hologram discussion. He called the entanglement effects, "spooky action at a distance."

He proposed a thought experiment in 1935 to demonstrate that the entanglement of particles was not true. It is called the Einstein-Podolsky-Rosen experiment, or EPR effects (Figure 4).

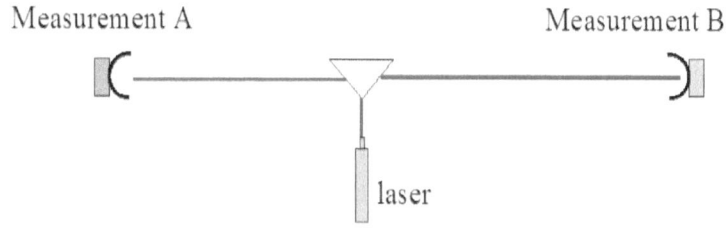

Figure 4 The Einstein-Podolsky-Rosen experiment

Physicist David Bohm proposed the simplest form of the EPR experiment. He suggested using the spin of electrons. Think about a spinning top or gyroscope. The object is spinning around some axis, so there is some angular momentum associated with the object. This angular momentum has direction related to the direction of the spinning axis. Electrons have a spin property that can be easily measured as the direction of spin.

Bohm proposed an experiment in which two electrons are created that together have a total spin of zero. When the two electrons are separated and are sent off in opposite directions, the spins of the individual electrons are entangled. This means that if you measure the spin direction of electron A, the direction of the spin of electron B will have to be 45 different from the value measured at A. In the experiment, you can choose which direction to measure after the electrons are separated.

Einstein argued that if the electron was local, then a measurement on A should have no effect on the measurement on B: there should be no spooky action at a distance. The experiment was finally performed in the 60s. Quantum mechanics won. There is "spooky action at a distance."

The point of all this is to demonstrate that for things with little or no mass, a single atom and smaller, the world is a very different place from the one you and I walk around in everyday. Matter and energy are in this amorphous, distributed, non-local, quantum state until they are measured. While they are in that quantum state, everything is everywhere. There is no separation between entangled particles, no matter what they look like in material world.

Einstein's problem, like many other people, was that he believed that two particles existed as separate entities from the time they were sent off in opposite directions until they were measured, like two cars leaving town, one heading east and one heading west. They both exist while they are traveling down the road.

Quantum mechanics describes a very different kind of reality. In a quantum reality, two entangled electrons do not "exist" as separate particles until they get to their respective detectors. Prior to that the question of where they are has no meaning because they are everywhere they can be all at once. They are really the same electron while they are traveling to their respective detectors. Spooky.

Non-temporal experiments

If non-local experiments bother physicists, then non-temporal results will drive them right up the wall. A non-temporal effect means that the result happens before the cause. This appears to upset causality, which is one of the basic assumptions in the foundation of all Western science. The assumption of causality says that the cause of an effect must happen before the effect. The EPR expriment has been used to demonstrate entanglement across time as well as space.

In the non-temporal version of the EPR experiment, detector B is closer to the source than detector A (Figure 5).

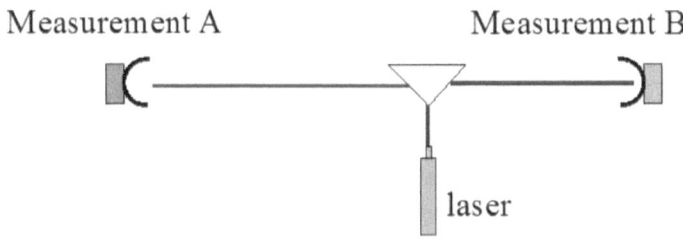

Figure 5 The non-temporal EPR experiment

That means the electrons will arrive at B before they arrive at A, so the measurement of the property at B will happen before the property is measured at A. From the first version of this experiment, we would predict that the measurement at B will specify what the measurement at A will be. And this is the case if we look at the results of the measurement at B. If we don't look at the measurements taken at B then something very interesting happens. When we look at results of the measurement at A first, and then go look at B, we find that the B result will be consistent with the A result, and not the other way around. Even though B was measured first. This implies that, somehow, the measurement at B was "changed" by looking at the result at A. Actually, that's a rather materialist view. A better way to say it is that the measurement was undefined until it was observed.

Influencing quantum events that "happened" in the past is quite common. Lynne McTaggart has reported on several in *The Field*.[3]

How can that be? This is a common question when people are confronted with things that differ widely from their normal experience. In most cases, I believe that a more accurate interpretation of those words is, "That can't be!" The answer to the first question -- how can it be? -- is that the quantum field hologram takes the Fourier transform of all four dimensions in space-time, including time. That means that time is as distributed and non-temporal as space is distributed and nonlocal in the quantum hologram. Communication can happen across time just like it can happen across space. The past is not fixed in the quantum field state.

Chapter 6   Reality and Quantum Weirdness

Actually, even using the word "communication" is a prejudice left over from Newton's local, material view of the universe. Communication implies sending some sort of message between two isolated bodies. The implication of quantum mechanics is that there are not two, isolated bodies. There is only one body that "knows" everything about itself.

Fortunately for those of us who live in large-scale material reality, these effects only happen at the quantum level, which means with things no bigger than a single atom, or a buckyball at most. By the time we get up to you and me and mountains, everything is properly local and the cause always happens before the effect. Later, we will see that our inner state is quantum-level information, which has very interesting implications.

All is one equation in the quantum world

We are talking about the weirdness of the quantum world. The experiments described in the previous sections are, indeed, weird, but, we can find plenty of weirdness in the basic nature of the quantum equations.

Quantum mechanics has demonstrated that the world at the quantum scale behaves very differently from the world that we see every day. I've used the terms, "distributed across time and space," and non-temporal and non-local to describe the quantum field. These terms are all commonly used in discussing quantum mechanics, but they are really a little misleading. They all imply the existence of separate things that are somehow spread around in the quantum field. I don't believe this is entirely accurate. I think the quantum field equations give us a better way to think about it. So, let's look at the quantum equations. I'm not going to show any equations. I'm just going to talk about them.

In conventional mechanics, we can write an equation to describe how an object will move, or evolve over time. If we have two objects, then we write two equations. If we have 10 objects, there are 10 equations. The situation in quantum mechanics is quite different. We can write a single field equation for a single, isolated electron. If we have two electrons, then they are both described with a single field equation. If we have 10 electrons, they are still described with a single equation. Those electrons are not discrete objects that somehow get tangled up, or entangled. In the quantum state, they are a single thing, whole and unified across time and space. There is no communication required, or even possible between the individual electrons because there are no individual electrons in the entangled, quantum state. It is only when they interact with something big, or they are observed, that this unified quantum field state collapses into an individual electron appearing at a specific place at a specific time.

Material existence and time

I have talked about how photons and electrons exist in time because their behavior can help us understand souls and health. We have seen how the quantum field equations allow things with little or no mass to be distributed across time and space. That distribution is what makes the quantum weirdness experiments work. But all of the individual photons and electrons that we actually observe are observed in one very specific place at exactly one specific time. We infer that they had to be distributed in time and space to account for the weird effects that we see.

Chapter 6   Reality and Quantum Weirdness

There is a name applied to the process of changing from the quantum-distributed state to a concrete observation. It is called the collapse of the wave function. It is a very awkward moment for physicists because there is no description in physics for how it happens. One moment things are waves, nicely described by the quantum field equations and their probability curves. The next moment a material photon appears exactly here on the film at exactly 10:29 AM, Eastern Daylight Time on April 23, 2008. It is embarrassing for physicists to not be able to explain how that "collapse" happens. It is, after all, where our material reality happens.

The idea of the "collapse of the wave function" is not only awkward, it is not accurate. The wave function does not collapse. It is still there after the first photon hits the film. In fact, it remains in place as long as the experiment is set up. All of the photons that go through the experiment are governed by the same wave function. Every one of those photons collapsed into a specific blip on the film all of which were bounded by the same probability curves.

What is the relationship between existence in a non-temporal, non-local quantum field, on one hand, and existence in a very local, material reality where clocks mark off a very linear time? This is a question that will be relevant to our discussion of souls, later in the book, but for now, let's stick with photons and electrons. What is the relation between the distributed quantum version of the photon and the very local, and temporal version that makes a specific spot on the film? I would like to suggest that the wave function of the photon continues to exist in the field after it has made its blip on the film.

How can that be? If the photon has changed from being a wave "before" it hit the film to being a light spot on the film after it collided with the film, then there can't be anything left some other place or time. This is the same problem that many people, including Einstein, had with the non-local quantum weirdness experiments. How can a single photon go both ways around the path and interfere with itself? The problem comes from thinking of the photon as a regular, material particle, like a marble. A marble can be here or there. It can't be here and there at the same time. The problem also comes from thinking about reality in terms of one equation for one particle.

To resolve the problem we have to recognize that the photon is not a material object. It is a quantum object. In its quantum state is it everywhere it can possibly be, all at once. This is not just an abstract idea. The photon really is everywhere and we have a lot of experimental observations that require that the photon be, literally, everywhere at the same time.

It also helps to recognize that the wave equation for one electron that we see on a film is not serving just that one electron. It is serving all the electrons. That means that the equation continues to "exist" even after one of its members has collapsed into a material blip on the film. The essence or the behavior tendencies of the electron continue to exist after the material electron has been absorbed into the film.

I said earlier that the quantum field does the same thing to time that it does to space. Things with little or no mass in the field are non-temporal. That is, they exist at every time, all at once just like a non-local object exists at every place, all at once. That means that our photon, even though it has collapsed into a blip on film at a

Chapter 6   Reality and Quantum Weirdness

specific time, still exists in the quantum field as the wave it was before it collapsed on the film. I think our sense of linear time is more deeply ingrained in our experience than our sense of place, which makes the idea of "everywhen" harder to grasp than "everywhere." For example, as I write this, my spelling checker is complaining about "everywhen," but "everywhere" is perfectly OK.

Actually, I don't think that the idea of everywhen is entirely foreign to lots of people. Many readers are already comfortable with the idea of something having both a timeless aspect and a material, time-limited aspect: our material bodies and our immortal souls, for example. I use the term non-temporal because it is more scientific and less religious than immortal, but they mean the same thing. In many spiritual belief systems, the soul exists independent of the material body, or bodies if you include reincarnation. My suggestion here is to take that idea and apply it to photons and electrons. This is why the non-temporal quantum weirdness experiments work. The electron still exists in the field "after" it has been recorded in material reality.

I understand that this line of reasoning may bother some people. Do electrons have souls, or do humans just have a non-temporal aspect that exists across time? Personally, I am delighted to find a unified model of existence. I work the same as electrons. Of course, I have a much more interesting personality than electrons. However you choose to think about it, we all exist outside and beyond the bounds of our normal, material time and place.

Life and Spirit in the Quantum Field

What good is all this?

I've given all of these examples of photons and electrons in the quantum state to demonstrate the nature of the quantum field for things with little or no mass. Later I'm going to suggest that the non-material aspects of living things are also distributed, or entangled, in the quantum field because they are also massless entities. The distribution and the entanglement are the same in both cases, but I'm not suggesting that electrons or photons entangle living things. This is a suggestion that has been made by other authors and I don't think it is necessary or valid.

The scale problem

When we talk about photons and electrons in the quantum field we use words that sound a great deal like a Taoist or Buddhist talking about the spiritual world. People have noticed this similarity from the beginning of quantum mechanics. Recall Niels Bohr and his family crest. The parallels between the two are striking and they are certainly very appealing. But there is a problem: Quantum effects only occur in things with little or no mass and all living things are very massive compared to photons, electrons and even buckyballs (60 carbon atoms). A single cell in an animal can contain 1,000,000,000,000 carbon atoms. Living things, even viruses, are huge compared to the scale where quantum effects are active. If we are going to use the quantum field to account for the non-material phenomena that humans and other living things experience, then we have to find a way to connect the macro scale of biological systems to the micro scale of the quantum field.

Chapter 6   Reality and Quantum Weirdness

Up until relatively recently, it has not been possible to make this connection. The basic assumption of the life sciences has been that there can be no quantum effects in living systems, so there is no need to even consider them in the life sciences. This is changing as the quantum revolution creeps into biology. We will look at the connection between material living systems and the quantum field in the next two chapters.

How far have we come?

For most readers, these last few chapters have been large steps.

We started with those 3-D pictures that most people are familiar with, holograms. I explained the Fourier transform, which almost no one is familiar with. The Fourier transform is important because it is the math that explains why we can have a non-temporal and non-local quantum field. To put that in more accessible language, it explains why spirits are eternal and why god can know everything. Actually, a better way to put that is to say that god is all knowledge.

We talked a lot about deterministic science and probabilistic science. Quantum mechanics and the predictions of quantum equations are definitely probabilistic and not deterministic. This bothers physicists. It is important to our journey because later it will account for why gods are not the all-powerful beings they are often portrayed to be. It also accounts for why humans can be god -like.

We saw a hint at the end of this section that "all is one" might be literally true because of the way we write the quantum equations. We also saw the challenge we will address next: conventional quantum mechanics only applies to things with little or no mass, like photons and protons, while we humans and other living things are very massive. We will lay some ground work in the next chapter and then connect living thing and the field in Chapter 8.

---

[1] M Arndt et al. 1999 *Nature* 401 680.

[2] Penrose, Roger *The Road to Reality*, Alfred A. Knopf, New York, 2004. Page 603.

[3] Mctaggart, Lynne, *The Field*, HarperCollins Publishers, New York, 2002, Page 171.

Chapter 7
Holograms in the Brain

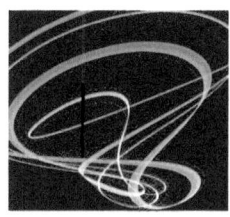

PHYSICISTS HAVE AT least acknowledged that there is some similarity between their description of how low-mass particles work and the Taoist descriptions of reality.

This is not too threatening to scientists because no one is suggesting that electrons have souls or communicate with spirits (although I seem to recall that I might have implied just that in the last chapter).

The life sciences (biology, botany, medicine, etc.) are in a more awkward position. On one hand, they have tried very hard to be properly scientific in Newton's terms. The life sciences can be characterized, I believe, as the study of molecular level chemical and electrical interactions in living systems. But the life sciences are concerned with living things, including humans, and lots of people do suggest that humans have souls and communicate with spirits. If you get too high up the function ladder of living things, you start to encounter phenomena that are very difficult to deal with in Newtonian terms. So restricting the study to molecules is safer.

On the other hand, even at the molecular level, there are many phenomena in living systems that are very difficult to account for with Newtonian science. This is the domain of quantum biology.

Life and Spirit in the Quantum Field

The normal approach of western science, both Newtonian and quantum, is to begin at the bottom, where you can solve at least a few equations. I am interested in the high level functioning of humans and other living systems, so solving equations is not an issue. At high levels of human function, the only observations we can make involve verbal descriptions of the inner experiences. The only instruments are human observations. At this level of functioning, we find ourselves in the same position as the Taoists and Buddhists. But we do have some good old Western science to guide us in our descriptions of that high-level functioning.

Where are we? In the previous chapter we looked at the holographic nature of the quantum field. We saw that information can be distributed across time and space relative to our state here in the time domain of large-scale, material reality. This property of the field is very appealing because it appears to offer a medium or mechanism that might be used to help account for the human-level phenomena that operate across time and space. There is, however, a problem with actually using the quantum field for processes at the scale of living systems.

The problem is that the things we know to behave in a quantum way are small. Human beings, and even human cells, are huge compared to those quantum objects.

Quantum mechanics is quite explicit about the effect of mass on behavior. The equations tell us that as the mass goes up, things become increasingly local and temporal (that is, they are no longer everywhere and everywhen). The largest particle that has demonstrated quantum superposition in the two slit experiment is a buckyball, with 60 carbon atoms. A buckyball weighs about 10^{-21} grams. Everything in living systems is big. A virus, which is just barely a living system, can be 100,000 times bigger than a carbon atom. A single cell contains 10^{13} atoms and weighs a nanogram (10^{-9} grams). A cell consists of mostly water with molecules of fats, proteins and carbohydrates. A single fat, protein or carbohydrate molecule can contain millions of atoms. So living systems, even the smallest pieces of living systems, are well past the size where quantum effects are directly possible.

Another problem is isolation. Quantum superposition, having photons be everywhere at once, requires that the particles be isolated from other particles. As soon as the particle bumps into something else, the quantum field equation (the wave state) collapses into a material instance of the particle. Living cells are very busy places. Most of the weight of an individual cell is water molecules, which only have three atoms each, so there are lots of water molecules in a cell. The biological molecules are much larger and make up a small fraction of weight, so there are very few of them relative to water molecules. The result is that water molecules are constantly bombarding all the molecules in a cell. I have heard the argument against quantum effects in living systems stated like this: Cells are too warm, wet and noisy to allow any quantum effects.

In spite of the objections, there are many phenomena in living systems at all levels, from molecules up to human cognition, that are very difficult to explain without some sort of connection across distance.

Fortunately for us, there has been a great deal of work done on quantum phenomena in living systems in the last 30 years. Much of it, of course, is on the fringes of mainstream biological science. Fringe or not, I have selected some of the proposals and assembled them into a coherent explanation of how action at a distance works. As I have mentioned before, there is very little current science, fringe or otherwise, that deals with the higher human function, specifically spiritual and other non-material phenomena. To fill in the gaps, I have added my own speculations ("Theories" might be too strong a word). The result is a model that accounts for how the non-material stuff works and has a few interesting implications for how we might live our lives.

In this section, I would like to build the bridge between living systems and the quantum field. To do this, I need to do a couple things. One is to connect the thoughts, feelings and perceptions in our mind to the quantum field. The other is to describe how the quantum field can think, and therefore create, at the quantum level.

My bias toward my material body

When I talk about the quantum aspects of life, I realize that I have a fundamental bias toward the material body. While this is quite understandable since I spent my formative years thinking that reality was the material world, it is not consistent with the message that I'm trying to convey.

That message is that the quantum field is reality and our material world is simply a projection of that reality through our senses. I will try to explain how that works, but be aware that I will be starting from our material brain and working toward quantum field, rather than the other way around. That makes it sound as if the thoughts are originating in our material brain. I don't believe that's the way it works.

Non-local connections in living things

At the outset of this book, I described some of the non-material phenomena that make up the human experience that I want to account for. These are high-level behaviors of human perception and cognition. Human beings are not the only living things that experience this non-material connection. It appears to be a property of all living things. Actually, connection and influence through the quantum field is a property of all matter, but for our purposes I would like to give a few examples of non-material connections in living things other than humans. Let's begin with plants.

Plants

In the 1960s, Cleve Baxter connected his polygraph to plants and found that they responded to material and non-material inputs across time and space. The measured responses of plants looked like the responses of humans on the polygraph. They showed what looked like increases and decreases in stress. He demonstrated that they have a reaction to physical events like being watered or having a leaf cut. More interesting, for our purposes, were the responses to human intent to harm or help the plant. That reaction was registered whether the human was in the same room or in another country.

Now consider some low-level animals.

Life and Spirit in the Quantum Field

Social societies: termites, sponges

Rupert Sheldrake, the British biologist, has spent many years championing a very unpopular theory, at least in mainstream biology. He has proposed the existence of non-material morphic fields to account for form and structure in nature and for coordination of behavior across collections of individuals. He cites a number of very interesting examples of structure and behavior coordination that are extremely difficult to account for using conventional physical senses.[1]

We have many examples of social societies of animals that behave in a very coherent and coordinated way. Sometimes the coordination is so extensive that a colony of individual animals appears from the outside to be a single organism. Sponges appear to be plant-like things attached to the sea bottom. But if you pick a sponge and push it through a metal sieve, it disperses into a cloud of individual animals. These individual animals swim around in the water for a bit and then reform themselves into another sponge "organism."

The Portuguese man-of-war jellyfish appears to be a single animal with specialized parts for flotation, motion, catching prey, digesting food and reproducing. But all of the special parts are separate organisms. The jellyfish "organism" is a colony of individuals who are specialized in form and function.

The social insects like ants appear to be collections of individual insects. There are workers, drones and the queen. Individuals of each type appear to be a whole organism, but none of them can exist outside of the colony. The queen cannot feed herself and only the queen can produce eggs. The workers cannot reproduce, and cannot even feed themselves away from the nest. So it appears that the nest itself is the digestive system. Among workers, there is considerable division of labor. The workers function as hands and mouth for the nest organism. In the case of ants, the nest is the organism. It is the collective entity that can sustain life and reproduce itself.

Termites appear to carry this notion of a collective organism one step further. In hot dry areas, termites build nests that stand up on the ground and can be over six feet tall. The workers, of course, build the nests from sand held together with some excretion. The workers are blind, but they work in a very coordinated way. The nests show a consistent size and structure within a given species. The compass termites in Australia, for example, build nests that have a base that is long and narrow. The long axis of the base is always aligned north and south. That arrangement minimizes exposure to the sun during the hottest part of the day, at noon when the sun is in the north in the Southern hemisphere. It is interesting that millions of apparently individual worker termites can produce such consistent and coherent structures.

In Africa, the *Macrotermes natalenis* termites build very elaborate nests. They live on fungus that they grow on wood carried into the nest. The nests have a very elaborate air circulation system to remove heat and carbon dioxide from the nest and bring oxygen back in. The nests are built from covered passages that then become tunnels when more passages are built on top of the earlier passages. Workers on either side of the passage under construction build the walls and when the walls get high enough, the walls curve over to join at the top and cover the passage. Somehow the workers coordinate their efforts so that the two sides are the same height and join correctly at the top. Some of the channels are large, four inches in diameter, and are used as air ducts. The nests can be 10 feet high and take many worker lifetimes to construct. This kind of coordination is very difficult to account for if we assume that each individual insect is an isolated organism connected to its environment only via its senses.

The nature of that control is indicated in some experiments done by Eugene Marias on termites in South Africa.[2] He observed that when he made a large gash in the side of the nest, the termites would immediately begin repairing the damage. Many individual workers were involved working on both sides of the gash. The damage was repaired and the nest was returned to its original structure. He wondered how the individual blind workers coordinated their efforts across the very large space of the gap. To see if there was some kind of communication between workers he made a gash and then inserted a large metal plate in the center of the gash to separate the workers on either side. The metal plate had no effect on their building activity.

Then he knocked the entire nest down and deposited the rubble and the termites in four boxes set next to each other to form a larger square. The rubble and termites in each box were separated from the termites in the other three boxes. All the termites immediately began rebuilding the nest. The new nest was the same size as the old nest. The presence of the box walls had no effect on the rebuilding of the nest and the work was coordinated in spite of the physical separation of the colony into four boxes.

Chapter 7 — Holograms in the Brain

Looking for other means of coordination, he tried removing the queen. That worked. When the queen was destroyed, all building activity throughout the nest ceased. It appears that the individual worker organisms do not even think for themselves. The queen conceives the nest and directs the workers in every aspect of its construction and repair. The queen is located in a chamber at the base of the nest and has no direct means of communication with the millions of workers throughout the nest. This is very hard to account for if only physical means of communication are available.

These examples demonstrate collective behavior of groups of what appeared to be individual organisms. In many cases, direct physical communication is not possible. Even plants, which we don't normally think of as communicating very much, demonstrate awareness not only of the state of other plants, but of the state of human emotion and intention. Sheldrake has proposed morphic fields as the medium of the communication. What I would like to propose here is that Sheldrake's morphic fields are the quantum field. This is not an original suggestion on my part.

The so-called "lower" animals and plants can communicate across distance, even when separated by physical barriers. I want to explain how that works in humans and I will take that up in the next section. But please keep in mind that all living things have this non-material connection. It is the nature of quantum life.

Thought/feeling as holograms

My goal here is to describe how the macro level thoughts and feelings can connect to the quantum field in a two-way communication. I would like to do that in three steps. The first is to describe human thought, perceptions and feeling as a hologram having a form very similar to that of the quantum field. The next step is to describe how those macro-scale holograms in our brains can be connected to the quantum field. The third step describes a mechanism that allows us to extract the information we want out of the sea of information that is the quantum field. So, let's begin with thought as holograms.

Pribram's brain

Carl Pribram is an M.D. and a researcher of brain function. He has devoted a great deal of attention to the mechanisms of visual perception. It is this work that led to his holonomic brain function model[3]. His work on this model was advanced by conversations he had with Dennis Gabor, the inventor of the hologram whom we encountered earlier, and with David Bohm, a physicist who made many contributions to the development of quantum mechanics. Gabor was developing a model of sound perception that was very similar to Pribram's. Both the Pribram and Gabor models depend heavily on our old friends, Fourier transforms. I would like to begin by considering visual perception.

How we see

When we look out at the world around us, if our vision is working correctly, we see a smooth, consistent and detailed image of our environment. Everywhere we look, we see equal detail. There is a common sense sort of model for how vision works. It goes like this (Figure 1).

Chapter 7 — Holograms in the Brain

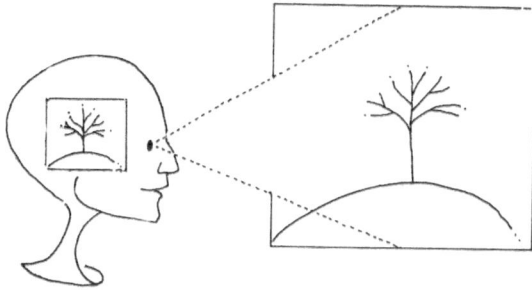

Figure 1 The normal model of visual perception

Light bounces off of surfaces in our environment and some of it enters our eyes. The lens in our eyes focuses an upside down image on the retina, which is full of optical nerve endings. The optical nerve carries the image up into the visual cortex where the brain figures out what it's seeing. Does that sound comfortable?

There are several problems with this model. The model implies that the eye works like a camera. It projects an image on to the retina and the retina captures the image. To do this, the retina would have to be a uniform light-sensitive device, like photographic film. One problem is that the retina is not at all uniform. Our vision has a small cone, about 5 , where we perceive sharp detail. That is not very big. If you extend your arm out in front of you and raise your index and middle finger together, that is the width of a 5 cone. Outside of that small central region, we have a large area that perceives mostly contours and gradations. And at the periphery of our vision, we see very little in the way of objects, but we are very sensitive to movement. That's left over from the days when we needed to detect the saber-toothed tiger sneaking up on us from the side. The bottom line is that our retina is not a uniform recorder of light like film.

Another problem with the retina is the hole. The retina is covered with optic nerve endings. All those nerve endings are collected at a point a little outside (toward the ear) of the center of the retina. Where all the nerve fibers come together there are no nerve endings, so there is a little hole in our retina. If you're not familiar with this hole, there is a small experiment you can do in the comfort of your own home to demonstrate its existence. Take both arms and extend them straight out in front of you. Make fists with both hands and raise your index fingers. Put your index fingers together so they are both pointing straight up and you are looking at the fingernails on your index fingers. Close your left (or right) eye and with your right (or left) eye look at the fingernail on your left (or right) hand. Keep your focus on your left fingernail. Then, without moving your focus, move your right hand slowly toward the right. Go about eight or 10 inches to the right and then slowly bring your index fingers back together. Repeat this slow, smooth motion several times. You may notice that at a point when your fingers are three or four inches apart, your right fingernail disappears. As you keep moving, it reappears. As you move your right hand back and forth through that spot, your fingernail will disappear and reappear. Your fingernail is over the hole in your retina when it disappears.

We can conclude, then, that our retina cannot record the smooth and continuous image that we normally perceive. There was another problem with the commonsense model that Pribram found: If the optic nerve carries a whole image up to the visual cortex, then there must be some place in the brain where the image is "displayed." Pribram went looking for that place using brain scans. He did not find it. Instead, he found that many different areas in the brain light up, depending on what is being looked at, but there was no one place that was always active. Facing a situation where the current theory was not supported by the observations, he did the scientifically appropriate thing and proposed a model that matched the observations.[4]

Seeing in holograms

He called his new model the holonomic brain theory. It works like this, Figure 2.

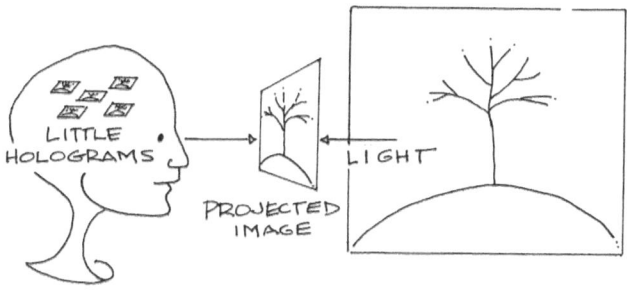

Figure 2 Holonomic perception model

Light is still reflected off the surfaces in our environment and collected by the lens of the eye. An inverted image is still focused on the retina. But instead of simply transmitting the image, the optic nerves do Fourier transforms on different aspects of the image. The central part of the eye does the transform on outlines, that is on the edges. The region surrounding the center does the transforms on contours and gradations. Recall that a

Fourier transform converts information from our material world into waves.

Back in Chapter 4 when we were talking about holograms, I said that the thing that made the hologram work was the presence of the phase information in the photograph. It appears that our optic nerves capture phase information from the incoming light[5]. The nerve endings are arranged in groups of three. Each triplet of nerves captures the phase of the light at that point. This means that our visual system can produce holograms without needing a reference beam.

So what gets transferred to the visual cortex is a collection of small holograms of different aspects of the image. Each of the different aspects is handled in a different area of the visual cortex. This accounts for the lack of a "central projector" in the brain. There is no image in the brain.

Pribram demonstrated this by showing subjects pictures that were visually similar but had different Fourier transforms and by watching brain activity. Changes in the activity corresponded with changes in the Fourier transform.

It is important that the light from our environment being received by our eyes is a Fourier transform. The lens of our eye does an inverse transform on the complete image and then our retina does a Fourier transform on different parts of the image. This means that our minds work in the Fourier or frequency domain.

Pribram suggests that we are continually taking these Fourier snapshots of different aspects our visual environment. All this processing takes time. This accounts for why discrete events that happen in less

Chapter 7 — Holograms in the Brain

than one tenth of a second appear to be continuous movement to us. It takes about a tenth of a second to take one set of snapshots. Our peripheral vision works faster than that. Have you noticed that you can see the flicker of a single fluorescent bulb in an otherwise dark room if you don't look directly at the light? A fluorescent bulb blinks at 60 cycles per second, which is faster than we can directly perceive. We can detect the flicker with our peripheral vision.

The perception process

Thus far, we have seen that Pribram is suggesting that there is no complete image in our brain, but we clearly see a complete image. So the question is, where is the image? Pribram's answer is that it is projected out of our eyes and we see it in front of us. The process works like this:

Little holograms from our eyes are delivered to the appropriate areas of our visual cortex. For each hologram, we go out to visual memory and retrieve a hologram that somehow matches the one we received from our eyes. (We'll see how holographic memory works later.) The collection of these familiar holograms is then sent back out the optic nerve and they are projected out of our eyes into the space in front of us. Ten percent of our optic nerves go out, that is from the brain to the retina. Ninety percent of the optic nerves go in, that is from the retina to the brain. The smooth, complete image that we are used to seeing is the result of interference between the light we are taking in from our environment and the image made of familiar elements that we are projecting back out.

This may seem like a rather cumbersome process, but it very nicely accounts for some common observations about human perception. When five people witness an accident, they are likely to give five different explanations about what happened. People are clearly not recording what they see like cameras.

Another feature of human perception is that we see what we are familiar with and we have a hard time seeing new things. The story is told about Darwin (which may or may not be true) sailing up the west coast of South America. When he put in to some of the harbors, it was the first time that the people living there had encountered a large ship. The story is that they did not see the ship. It did not register in their perception. When the crew put the long boats in the water to row into shore, the natives saw those because they were accustomed to seeing small boats.

An example closer to home is birdwatchers, or birders. I am not a birder, but I have been around a few. When I walk in the woods, I might see a cardinal and a robin and hear some birds that I don't recognize. The birder walking next to me would see and identify 10 species, including an obscure vireo, and identify the songs of eight others. We are both taking in the same light and sound from the environment, but because the birder has a different set of familiar images than I do, she perceives a very different set of things.

All the senses work like vision

Dennis Gabor[6] proposed a very similar model for the perception of sound. In his model, the sound waves are received from the environment and are converted into multiple Fourier transforms. We retrieve Fourier transforms that match and project those Fourier transforms back out through our ears. The continuous words or music that we perceive is the result of that projection.

Pribram's suggestion is that all of our senses work in this manner. Everything is converted into Fourier transforms, or holograms, before being processed in the brain. Having all of our sensory input in the same form successfully accounts for the well-established phenomena of interchangeable sensory processing. The best example of that is tactile input being processed in the visual cortex of blind people. Blind people who read Braille have neural pathways from their fingers into what once were the visual processing areas. Alvaro Pascual-Leone demonstrated that when sighted people were blindfolded for as little as two days they began to develop pathways from their tactile processing areas into the visual cortex[7]. For this to happen so quickly, the form of the information handled in the tactile areas has to be the same as the information handled in the visual cortex. Pribram has suggested that the common form is a hologram.

Our perception of sensory information is not a simple recording process, like a camera or tape recorder. It is an active process that wraps the raw inputs from the environment in things that are familiar, things drawn from sensory memory. We'll see in the next section that our perception of non-sensory, that is, intuitive information works the same way. Hold this thought. It will be very important when we talk about perceiving the information that people have always identified as spiritual.

Perceiving intuitive information

This model also does a very good job of explaining why people perceive intuitive information in so many different ways. People can read or know the state of another person (apparently) through different physical senses.

Clairvoyants may see the information as colors around the body of the subject, usually called auras. Other people get sensations in their hands. These are practitioners of Healing Touch, Therapeutic Touch, Cranial Sacral and Polarity techniques. Medical intuitives just think about the subject and report the state in words.

All those people are reading the same information from the same source: the quantum hologram. The information is not arriving through any of the normal physical senses but through our intuitive sense. I believe the process is the same as for normal sensory input: We take the incoming information and go out to sensory memory to retrieve something that is familiar. The retrieved information is sent out via one of the regular senses. Exactly which sense gets chosen depends on how each individual learned their skill and their natural preferences, but this is getting ahead of the story.

To summarize, Pribram's process for visual perception is that we do Fourier transforms of many aspects of the image that is focused on our retina. Those Fourier transforms are handled in different parts of the visual cortex. For each of the little Fourier transforms, we retrieve a Fourier transform from our visual memory that matches the one we got from outside. Those recalled Fourier transforms are collected and projected back out of our eyes to the space in front of us. The smooth, complete image we see is the result of that projection. This process is the same for all of the inputs from our environment. Remember that our environment includes the material world and the non-material world, so the process applies to perceiving intuitive, spiritual and all the other non-material parts of the human environment.

We think and feel in holograms

If our sensory perception is handled in holograms, then it seems quite reasonable to assume that all the rest of our mental activity is handled in holograms as well. This includes our thoughts, feelings, emotions, memories and ideas. Our entire inner life is conducted in the Fourier frequency domain. Here is the first step of our connection to the quantum field. Our inner life has the same form as the quantum field and it has the same relationship to the material world that the quantum field does.

How is this relevant to our journey? Pribram's main contribution to our journey is the suggestion that we think and perceive in a Fourier frequency domain. This is important because the quantum field is also a Fourier frequency domain. The fact that our thoughts/feelings and the quantum field have the same form makes it much easier to propose that our thoughts originate in the field and that the field thinks/feels. The other useful part is the step in the process where we retrieve holograms from our memory to build the image that we perceive. This will be very useful when we talk about how people perceive spiritual and intuitive information.

Where are the holograms in the brain?

My goal is to describe how our macro-scale thoughts and feelings are transferred to and from the micro-scale of the quantum field. Thus far, we have proposed that the content of our minds and the quantum field have the same form, but they work at very different scales. To describe the connection, we need to look in more detail at the mechanisms of perception and consciousness.

I would like to begin by considering where Pribram's holograms live in our brains. Our brain is made up of lots of neurons, some 35 billion of them. Neurons are nerve cells that have long extensions, called dendrites and axons, for connecting to other neurons. The conventional view is that our brains work through the action of neurons. Each neuron collects input signals from its dendrites, which are connected to the axons of other neurons. When the input signals reach a certain level, determined by the neuron's training, the neuron "fires" and sends a signal down its axons, which results in new inputs to other neurons.

Computers have been built using this model. They're called neural net computers. A neural net computer is not programmed, it is trained. The effect of the training is to adjust the level of input at which each neuron fires. These kinds of computers are very good at some tasks that are difficult for conventional computers, such as character recognition.

Neurons certainly work this way, but in Pribram's model they are not the main source of thought and perception. Pribram suggested that the holograms are carried in the synapses between the neurons. Actually, the synapses were one place they might be carried. He suggested others. For my purposes, having the holograms in the synapses works best.

The neurons in the picture, Figure 3, are not accurate in that it shows neurons having a dozen or so ends that can be connected to other neurons. In fact, individual nerve cells can have between 1,000 and 10,000 synapses each. The synapses carry electro-chemical

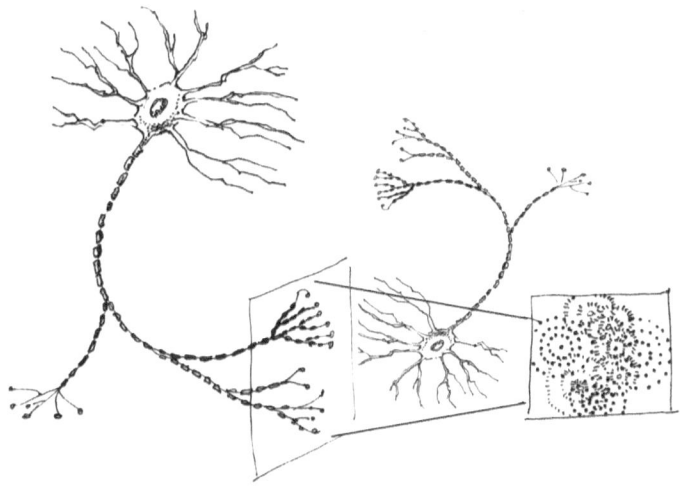

Figure 3 Neurons and synapses

signals between two neurons. Pribram has suggested that the voltage in each synapse oscillates and that groups of synapses have voltages that oscillate coherently and carry a hologram. With a thousand to 10,000 synapses per neuron and 35 billion neurons, there is room for lots of holograms in our nervous system.

Our first step

We have taken the first step toward connecting our inner state to the field: our inner state has the same form as the quantum field. It is holograms, or Fourier transforms, of the information we receive through our senses. All of our inner state - feelings, thoughts, emotions as well as sensory information - is carried in the frequency domain of the hologram.

Our next step is to bridge the gap between the macro-scale of living systems and the micro-scale of the quantum field. It helps that the waves that make up our inner state have no mass. They are just information.

---

[1] Sheldrake, Rupert, *Presence of the Past*, Random House, New York, 1988.

[2] Sheldrake, Rupert, *Presence of the Past*, page 230.

[3] Pribrahm, Carl, *Brain and Perception: Holonomy and Structure in Figural Processing*, Laurence Erlbaum, Hillsdale, NJ 1991.

[4] Pribrahm, Carl, *Brain and Perception: Holonomy and Structure in Figural Processing*, Laurence Erlbaum, Hillsdale, NJ 1991.

[5] Ibid, Page 62.

[6] Gabor, Dennis Acoustical quanta and the theory of hearing. *Nature*, 159(4044), 1947, 591-594.

[7] Pascual-Leone, Alvaro, Torres, Fernando Plasticity of the sensorimotor cortex representation of the reading finger in Braille readers, *Brain*, February 1993; 116: 39 - 52.

Chapter 8
Connecting to the Field

WELL, NOW WE have located the holograms in our neural tissue and they have the same form as the quantum field, but that does not address the question of how our the macro-scale holograms are connected to the micro-scale quantum field. For that, I want to turn to the work being carried out at The Center for Consciousness Studies at the University of Arizona. See http://www.quantumconsciousness.org/penrose-hameroff/consciousevents.html .[1] They have proposed that microtubes in our cells are the basic units of conscious awareness at all levels, from the individual cell up to human beings. Since microtubes are not a common topic of conversation I'll begin with the description of what they are and how they work in conventional cell physiology.

The microtubes

Microtubes, in conventional biology, are responsible for structure and motion in cells. They are made out of a protein called tubulin. Each microtube is a tube made out of rings of 13 tubulin molecules each, so each microtube is 13 tubulin molecules in circumference. The microtubes in a cell are continually forming and dissolving. While very few people are familiar with the term "microtube" almost everyone has seen pictures of them in action. Figure 1 is a picture of the chromosome in a dividing cell. The dark bodies on either side of the chromosome are centrioles, which are made up of microtubes. The little strings that connect the centrioles to the chromosome are also made of microtubes. The strings pull the two halves of the chromosome apart during the division process.

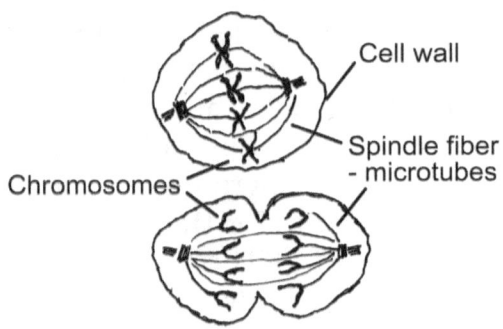

Figure 1 Microtubes in cell division

What's the connection between microtube and neurons? It turns out that neurons can move. They can pick themselves up from one synapse and put themselves

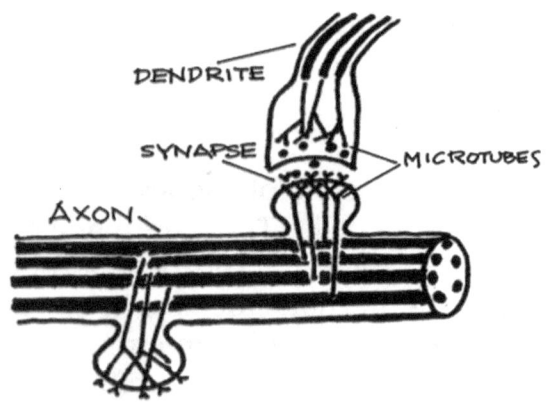

Figure 2 Microtubes in neurons

Chapter 8 — Connecting to the Field

down on another. This means that there are lots of microtubes in neurons, Figure 2.

And now, you might ask, how does that help connect our macro scale holograms with the micro scale quantum field? To answer that, we have to look at the quantum behavior of microtubes and their role in consciousness.

Tubulin is a protein and like all proteins, it is a long chain of amino acids. How a protein functions is dependent on how that long chain is folded. Every protein has one or more specific fold configurations that allows them to carry out their function in the cell. Tubulin has two fold configurations, shown as black and white in the figure. The proposal from Arizona is that the fold configurations play another role in the cell: The pattern of fold configurations, the pattern of black and white in Figure 3, represents an instant of conscious awareness. When the microtubes are in neurons, the consciousness can be our regular human consciousness.

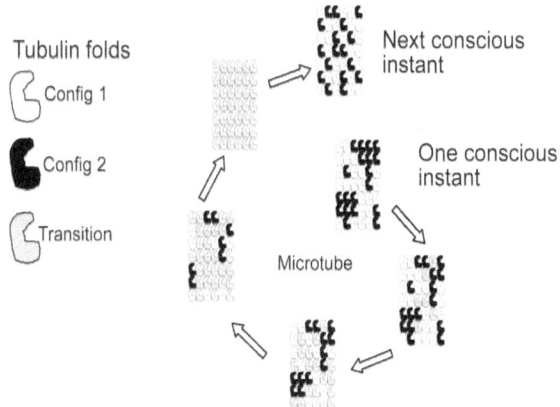

Figure 3 Tubulin fold patterns in microtubes

The process that the microtube uses to move from one conscious state to the next is critical to our journey. In Figure 3, the pattern of black and white tubulin molecules represents the conscious state of the microtube. The gray tubulin represents a molecule in the quantum indeterminate state, where it is in all the possible configurations it can be, all at once. The microtube begins with all of its tubulin collapsed in a state of consciousness. All of the tubulin molecules are in one configuration or the other. Then a few tubulin molecules slip into the quantum state, and then a few more, and a few more until all of the tubulin in the entire microtube is in this quantum, indeterminate state. And then, under the influence of what the folks at Arizona call quantum gravity, all of the tubulin in the microtube collapses into a new, material pattern of black-and-white configurations. This is a new instance of conscious awareness. Shortly afterwards, the process begins again as a few of the tubulin molecules flip into the quantum indeterminate state.

The important thing is that the transition between states is a quantum transition. Let's look at what that means.

Quantum behavior of microtubes

The reason this is interesting for our journey is the way a microtube moves from one instance of consciousness to the next: It is a quantum transition. Here's what that means. Each tubulin molecule has two fold configurations. Like many proteins that can move from one configuration to the other, that transition is a quantum transition.

Let me demonstrate what this means with an example using me. Imagine that I'm standing here in front of you and that I want to move over there, and "over there" is three feet away (Figure 4).

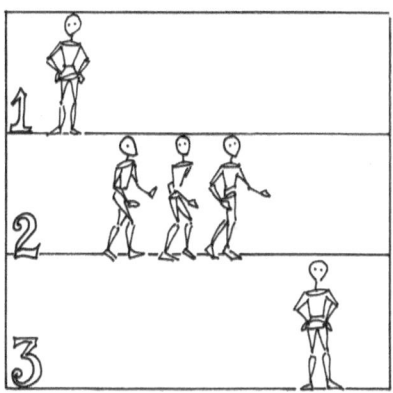

Figure 4 Material transition

I start to move and shuffle slowly toward my target. Say it takes me two seconds to move from here to there and you watched me the whole time.

I could ask, "Where was I after one second?"

Since you saw me move, you could confidently say that after one second I was halfway between here and there. I could then ask, "Where was I after 1 ½ seconds?" and, you could confidently answer that I was a little more than two feet away from here and a little less than one foot away from there. The thing that makes this a non-quantum transition is that we all knew exactly where I was at every moment during the transition.

Now, let's repeat this little process, but this time I'll be a quantum object and the transition will be a quantum transition, Figure 5.

Life and Spirit in the Quantum Field

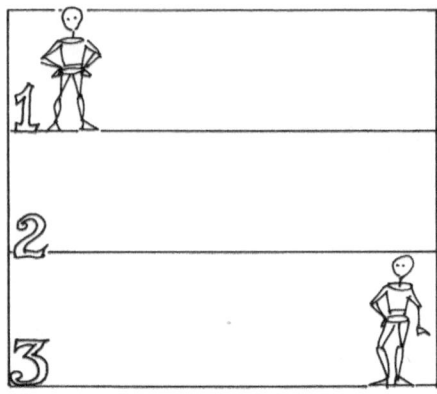

Figure 5 A quantum transition

So, here I am standing here and I'm going to move to over there. I start to move and the first thing that happens is I disappear. The next thing that happens is that two seconds later, poof! I appear over there. If I were to ask you where I was after one second, you couldn't say because you couldn't see me. The question, "Where was I?" has no meaningful answer in the quantum state. The only answer I can give is that during the transition time I was everywhere I could possibly be all at once. When I enter quantum state I am distributed, or non-local. When I appear at my destination we say that the quantum state, or the quantum field equations, collapse into a specific physical state. It is this indeterminate state during the transition that makes it a quantum transition.

The warm, wet and noisy problem

Any interaction between an object in the state of quantum coherence and another object causes the quantum coherence to collapse. The interiors of cells are very busy places, so it appears that it would be difficult for the microtubes to be isolated from the noisy environment long enough to make this leisurely oscillation between the quantum coherent state and the collapsed state possible. The folks at the University of Arizona have suggested two possible mechanisms of providing that isolation.

One is a "shell" of water molecules around the microtubes. Water molecules are polar, that is they have one end with a slight positive charge and the other end has a slight negative charge. This polarity accounts for why water has such a high boiling point. The microtubes have a slight charge and they can attract a shell of water molecules. This shell might provide enough isolation needed to allow the quantum coherent state to exist.

The other suggestion is that the fluid that fills the cell's interior, the cytoplasm, has two states. One state is the normal fluid gel and the other is a fairly rigid crystal. The cytoplasm is capable of switching between these two states very rapidly. If everything was held in one place in the crystallized cytoplasm, the microtubes could have enough time to enter or exit the quantum coherent state.

So, while cells are warm, wet and noisy, there appear to be mechanisms in place to provide the isolation the microtubes need for quantum behavior to exist.

With those explanations out of the way, let's look at how all these quantum transitions help us think in the field.

Life and Spirit in the Quantum Field

Getting information from the quantum state

If this process seems a little bizarre, it might help you to know that what I have just described is a quantum computer. The people who do research on new computing technologies are working very hard on quantum computers, so cycling between quantum distributed states and collapsed, material states is not entirely unheard of.

In conventional computers, the basic operating unit is the bit. A bit can have a value of one or zero. In a quantum computer the basic operating unit is a quantum bit, or a qbit. In a quantum computer, there is no programming. The problem is entered and then the computer, that is all of the qbits, go into the quantum indeterminate state where they are in all possible outcomes at the same time. Then they collapse into one of the possible solutions. A big challenge for the makers of quantum computers is finding suitable qbits. Not only do they have to be small enough to display quantum behavior, we have to be able to read their state. Tubulin might be a fine qbit, but it is very difficult to read its state electronically. Living systems do not seem to have any trouble doing so.

The quantum computer work does demonstrate that objects in a quantum distributed state can retrieve information.

Connecting with the microtubes

In tubulin, we have a quantum object, a quantum computer, really, floating around in the macro scale of cells. I would like to propose that it is the microtubes

Chapter 8   Connecting to the Field

that provide the connection between our macro scale mind and the micro scale of the quantum field.

This little conjecture is my contribution to the progress of humankind. With this conjecture, I can account for a great deal of the human experience that is difficult for current science to handle. Let's consider how this might work.

We return now to the synapses in the brain with their oscillating voltages carrying the holograms of our perceptions, thoughts, feelings and emotions. The neurons on either side of the synapse are full of microtubes. The microtubes are alternating between concrete, material configurations of their tubulin building blocks and a distributed, quantum state.

The question of the moment is: How do the microtubes connect our macro-scale holograms in our macro-scale neurons to the micro scale of the quantum field? So, here are my suggestions:

1) The microtubes are quantum bits or qbits. Actually, they could be more than binary bits. They could easily have more values than just two. But the important thing here is that microtubes alternate between representing macro-scale information in the pattern of fold configurations of the collapsed tubulin molecules (all the tubulins are either black or white in Figure 3), and being in a state of quantum coherence (all of the tubulins are gray in the figure). They can carry information between the macro scale neurons and the quantum field.

2) The microtubes near the synapses in neurons are immersed in the oscillating electric field that carries the hologram. So, they will have access to the "information"

carried by the hologram. A single hologram in the brain will occupy many synapses

3) The microtubes near the collection of synapses that carry a single hologram are entangled so the microtubes are collectively "aware" of the complete hologram.

4) When the microtubes cycle into the quantum coherent state, they carry the information in the hologram into the quantum field.

5) When the microtubes cycle back into collapsed, macro state, they can carry information from the field into the hologram carried by the synapses. Communication is two-way.

The microtubes are responsible for transferring the information from our macro-scale holograms to the quantum field. They are also responsible for transferring information from the quantum field hologram into the macro-scale holograms of our mind. They do the same thing at all levels of living entity, all the way down to single cells.

Non-local, non-temporal thought

If the microtubes transfer information between our macro-scale cells and the quantum field, where does it "go" in the field? Here's another of my proposals: The information that drives the behavior of individual cells, as well as all of our mental holograms, is pure information. Information has no mass. When the information is transferred to the quantum field, it can be distributed across space and time like all other mass-less objects. Perhaps "transferred" is not quite the right word, since it implies being moved from one place to

Chapter 8 — Connecting to the Field

another. I think a better word might be "translated." That implies changing form, but not necessarily changing location. Nothing has to "go" anywhere. Everything is "right here."

There is also a problem with the word "information." It is correct to say that living systems run on information at all levels, from how to make a protein in a cell up to what to wear for that hot date. The problem is that the word "information" implies a static, and certainly non-living kind of thing, and that is not what I am talking about. The information in the field is living, moving and changing.

The material bias, again

We like to think of our material bodies as being real. In the normal view, our bodies, including our brains, are the source of the movement, choice, thought and feeling that make up our existence. The reasoning goes, if the quantum field is real, then it must be some kind of shadow of our "real" material existence that follows our material movements like my light shadow follows my material body. I believe that it is the other way around, as I have said before. The source of the movement is the quantum field. Our material experience is a projection of the information in the quantum field.

I find my bias toward the reality of my material body runs very deep. In the opening paragraph of this section, I said that the information of the thoughts and feelings are "translated to the quantum field." That makes it sound like they started out in our material mind and got put in the quantum field. I believe it is more accurate to say that the thoughts and feelings originate in the quantum field and get projected into our material minds. Just like the action of quarks and electrons: we have to calculate what happens in the quantum field and then see how that is projected into the material world.

The conclusion to be drawn from this mechanism is that our thoughts, feelings and the rest of the non-material aspects of our beings originate in the field and are both non-local and non-temporal. That is, they are everywhere and everywhen. Now we have a mechanism that can account for all of the non-local phenomena that are so troubling in the Newtonian view of the universe.

I should also point out that if this is the way the universe works, then it works this way all the time, not just when doing remote viewing or medical intuiting. All of our thoughts and feelings, including the most mundane and materialist, are distributed everywhere and everywhen.

All living things do this translation

In the description I just gave, I was talking about human thoughts and feelings being transferred to and from the field. But at the opening of this chapter I gave some examples of nonmaterial communication between nonhuman living things, like termites. I believe this same mechanism applies to all living things, beginning with single-celled living things, continuing up to multi-cellular living things, like termites and human beings and continuing up to the highest levels of human thought and perception. It is also the source of intelligence at the cellular level of multi cellular organisms. This is the mechanism for nonmaterial communication in all living systems.

Chapter 8 — Connecting to the Field

The mind as a quantum device

A profound implication of this model is that all of the activities of the mind are quantum phenomena because they all rest on the quantum behavior of the microtubes. Our thoughts and feelings are carried in the quantum field as quantum-level (that is, mass-less) information, so they are subject to influence from other quantum-level forces. We will look at the mechanism for that influence in the next section on retrieving information.

Memory in the field

Memory has been a long-standing problem for brain researchers. The commonsense view is that memory is held in the brain. The brain thinks and feels and it can talk about memories, so the memories must be in the brain. The problem is that when people have tried to find the location of memory in the brain they have not been successful. Experiments with salamanders and rats have shown that the animal can successfully remember previously learned things even when the brain has been severely damaged and disrupted. So, if memory is not in the brain, where is it? [2]

Based on the mechanism proposed here, memory is in the field. That's not an original idea, of course. Several people have made that proposal.

I like to use a television analogy. If you look at a television, you see all those little people walking around, talking, singing and selling pharmaceuticals. All those people are obviously inside the television, right? I can prove that by taking a screwdriver and poking around inside the television. When I damage something in the television, I damage the little people. So the little people are in the television.

We know, of course, that television is simply a receiver of information. Damaging the television damages its ability to receive and display the information. The brain is also a receiver, and a transmitter, of information. Or perhaps I should say the brain is a translator of information contained in the field.

If memory and lots of other things about my inner life are in the field, how do I retrieve them?

Holographic memory

I have described the field that holds all of these memories and emotions as one big entity. I assume this is the case because it appears that anyone can access anything else. For example, a medical intuitive needs only a name and a city in order to "connect to" that person. So if the field is one giant hologram, then it is interesting to look at how this retrieval can work. How is it that when you and I remember our memories of childhood vacations, I get our family trips to Michigan while you get your trips to the Jersey Shore? The field must be very crowded with information. Crowded or not, medical intuitives and many others can access information about other people if they choose to.

Once again our friends in computer research provide a mechanism that can account for this amazing retrieval. This time the topic is holographic computer memory. I mentioned earlier that holograms can carry a great deal of information, which makes them attractive for use in computer memory. But they have an even more desirable feature for use in computers: content addressable memory. I'll describe what that means in a minute.

Instead of using film as I described in our previous hologram discussion, holographic memory researchers are using a cube of photosensitive material 1 cm on a side, or about half an inch, Figure 6[3].

As before, a laser beam is split with one half forming a reference beam and the other half forming the object beam. The object beam goes through an LCD screen, or spatial light modulator, just like the screen used in

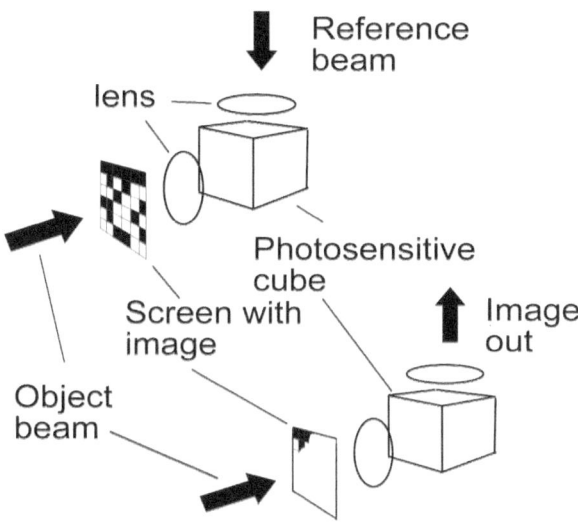

Figure 6 Holographic computer memory

computer projectors. Any kind of information can be put on the screen, from pictures of your trip to Niagara Falls to patterns of little black and white squares representing binary data. So the object beam, with an image of the information being stored, shines on one side of the cube.

The reference beam shines on another side of the cube and it hits the cube at an angle that can be changed. This records one hologram in the cube.

If the angle of the reference beam is changed, another image can be projected in the object beam and recorded as a hologram in the cube. When I first read about these experiments, they had been able to put 1,000 images in a single 1 cm^3 cube. When I looked up the subject again, I found that they were able to put 5,000 images in a single cube.[4]

Content-addressable information

Since this cube is intended to be used as computer memory, it is important to be able to get the information back out. There are two ways of doing that. One way is to use the reference beam to project each of the images. If the reference beam shines on the cube at angle one, the image that comes out is the image that was recorded at angle one. If the reference beam is moved to angle five, the image recorded at angle five is produced. That's the way regular computer memory works and it's not very interesting for our purposes.

What is much more interesting for us, and for the computer researchers, is content addressable memory. Content addressable memory refers to the ability to retrieve something from storage using just a fragment of the content you want to retrieve. In conventional memory, like computer disks, if you have a long list of the addresses of policyholders in your insurance company and you want to find the address of Joe Smith, then you have to start at the beginning and pull out each record and see if the name on the record is Joe Smith. When you find a record with a name that matches Joe Smith, then you can read that record which will include the name and the address. Actually, computers these days are more sophisticated than that, but you still have to open many records to find the one that has the content that you're looking for.

With content addressable memory, I can directly and immediately open the record, or records, for Joe Smith. This would save lots of time. Google already does large searches very quickly. Content addressable memory would allow Google to produce their search results with much less computer power than they are using now. The appeal of holographic memory is that it provides content addressable access to the stored information.

To do content addressable access in our holographic cube, the object beam is used instead of the reference beam. A small fragment of one of the stored images is displayed on the LCD screen (lower part of Figure 6). When the object beam is shown through the screen, the image that is projected from the cube is the image that matches that small fragment. What I find amazing is how much of the original image is required in order to retrieve the original image. They only have to put 0.05% of the original image on the screen in order to retrieve the complete original image. For those of you who are a little rusty on your percentages, that is 5/10,000 of the original image, a very small piece of the original. If the fragment that you put on the screen matches several original images, then all of the matching images will be returned.

Holographic memory is a relatively new area of research. I find it amazing that they can put 5,000 images in a 1 cm cube and retrieve one of them with such a tiny fraction of the original as a clue. Nature, on the other hand, has been working on this for several billion years. Nature has also demonstrated that she can be very subtle in her uses of information. So it would not surprise me if living things can retrieve their own information in the field with even less than 0.05% the original information.

I would like to propose that the big hologram in the sky, the quantum field, works the same way as the holographic memory in the lab. The thoughts, feelings, emotions and memories of all living things are routinely and continuously translated between the macro, material state and the quantum field. Of course, not all living things have thoughts like humans, but everything has feelings, emotions and memories. The bottom line is that all of the information about all living things exists simultaneously in the field. We are all able to access our own stuff and avoid other peoples' stuff because of the way holographic memory works. We only need the tiniest fragment of information unique to each of us in order to return our own unique information from the quantum soup of information. With this mechanism, we can understand how some people can access the information about other people. Medical intuitives are a high-level example of this behavior. They only need a name and a city to access the complete mental, emotional and physical state of that person.

If the quantum field is "like" holographic computer memory, then we have the ability to put things in as well as take things out. The new memories, thoughts, feelings and intentions that we have are all written into the quantum field using the same machinery that we use to retrieve the information, the microtubes translating holograms between the macro holograms in our brain and the micro scale of the quantum field.

Chapter 8   Connecting to the Field

Old habits die hard

Once again, I find that I have slipped into the point of view that the material world where we live is the source of movement and action and that the quantum field is a passive recorder of what happens in the "real" material world. I don't believe that this point of view is correct, but our language and all of our everyday conventions of thought were developed from that point of view. The new reality appears to be that the impetus for movement and the movement itself all originate in the field. Our material macro-scale reality is really a projection of the underlying <u>real</u> reality in the quantum field.

Holograms and holomemory are fine examples of how the field works. They are material things that we can draw pictures of and talk about, but they are material things that do not move. They are fixed and static. In that respect, they are not good examples. The field, including the inner states of all living things, is living, moving, growing and changing. Our material life, including the non-material thoughts/feelings in our material brain, is a moving picture show projected and filtered from the living field. It's a hard idea to grasp after all these years of living in Descartes' world.

How far have we come?

We have accomplished a very worthwhile thing in the last couple of chapters. We have connected our large scale mind to the micro-scale of the quantum field. Along the way, we saw that Pribram's perception model describes a very active, creative process for building what we perceive in the world around us. This is why our models of reality that I talked about in the opening chapters are the way they are. All of our perception is built from things that we remember, things that we learned.

We took a very important step toward our lofty goal of describing god and spirit. Our thought and memory have the same form as the quantum field and they exist in the quantum field. Given the nature of the field, we have to say that our thought originates in the field. That is our first step toward a field that thinks on its own.

All of this connection with the field is made possible by the microtubes in our cells. They are quantum computers that oscillate between the material world and the distributed quantum world. The other reason we can connect with the field is that our thought and feeling have no mass. The rules of quantum behavior can be applied to them.

I took the time to remind myself, again, that the field is not some shadow that follows my material body. Nor is it something separate that I have to "connect to". It is the origin of reality. I am a projection of information living in the field, not the other way around. This idea will become increasingly important as we travel down our path.

Now we need to consider the body. I said back in Chapter 2 when we were talking about the nature of religions, that the head or mind-centered religions were neglecting an important part of our being: the body. Up to this point our focus has been the brain and mind, but they are not the whole story, so let us next consider how the body works. We will see that it, too, is a quantum process.

---

[1] Hameroff, S.R., and Penrose, R., (1995) Orchestrated reduction of

quantum coherence in brain microtubules: A model for consciousness. *Neural Network World* 5 (5) 793-804.

[2] Lashley, K. S. In search of the engram, *Symposium of the Society for Experimental Biology*, 4:454-482, 1950.

[3] Moser, C, Psaltis, D, Holographic memory with localized recording, *APPLIED OPTICS*, Vol 40, No. 29, 10 August 2001, pg 3909-14.

[4] Li, Phillips, Hesselink, McLeaod, Associative holographic memory, http://doi.ieeecomputersociety.org/10.1109/AIPRW.2000.953606.

Chapter 9
Life Is Light

THE QUANTUM REVOLUTION is coming to the life sciences. As we will see, the result is that our health is a quantum process, and therefore is subject to quantum level influences and entanglements. Health, I should also point out, is usually associated with the body, so in this chapter, we are going to connect our bodies to the field. I think the heaven and head-centered religions are missing an important part of the spiritual experience: the body.

We are going to say that life is light and the light is in the quantum wave state. Life is a quantum process. We'll see in later chapters why this connects us to spirit. For now, let's consider the nuts and bolts of light in the body.

Biophotons

There are two kinds of light we know about in the body. Actually it's not visible light, so two frequency ranges of electromagnetic radiation would be more accurate. I'll begin with biophotons.

The first signals

The Russian scientist, Alexander Gurwitsch, working in the 1920s was the first to report emissions from a plant that could affect another plant[1]. He suspended one onion plant vertically and aligned another horizontally with its root pointing at the vertical plant, on the left in Figure 1.

After three hours' exposure, he examined a cross-section of the vertical root. He found that there had been 25% more cell divisions in the area exposed to the "rays" from the other plant, on the right. To discover the nature

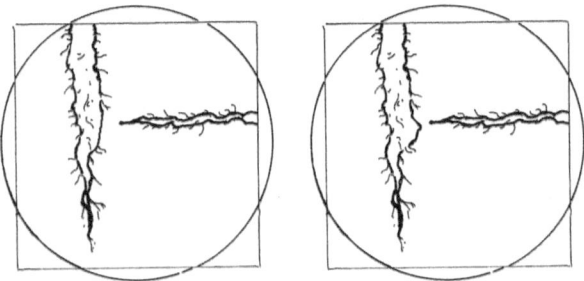

Figure 1 Gurwitsch's experiment with onion roots

of those rays, he tried inserting several different kinds of barriers between the two roots: plain glass, plain quartz and quartz covered with gelatin. Everything but the plain quartz blocked the effect. If the rays were electromagnetic, that indicated that they were ultraviolet waves or shorter. Quartz is transparent to ultraviolet light while glass and gelatin are not.

Gurwitsch's work caused quite a stir at the time. Many people tried to replicate the results. Some were successful and some were not. As a result, his work was relegated to the category of fringe science. Tompkins and

Bird suggested in T*he Secret Life of Plants* [2] that the effect may be influenced by human intention. We'll talk more about human influence in a later chapter.

Measuring the effect

In 1970, Fritz Popp was a biophysicist at the University of Marburg in Germany. He was studying the absorption spectra of chemicals that cause cancer. It is well-known that organic chemicals absorb specific wavelengths of light. Each chemical absorbs a different set of wavelengths. This is the basis for the practice of spectroscopy which is used to identify chemicals by their absorption patterns. Popp noticed that chemicals that cause cancer all seem to share an absorption band at 380 nm wavelength (a nanometer, nm, is one billionth of a meter). That wavelength is in the range of ultraviolet light. Absorption at 380 nm appeared to be a very good predictor of whether the chemical would cause cancer. [3]

There were some other connections to ultraviolet light. Sunburn is caused by the ultraviolet light in sunlight. If we expose our skin to the sun for too long, the ultraviolet light begins to destroy the cells. 380 nm is in the UVA range of ultraviolet light. UVA light contributes to sunburn. Popp was also aware of a phenomenon called photorepair. This is a phenomenon that is not understood, but is widely reproducible. Damaged skin, even severely burned skin, can be quickly healed by shining a very weak ultraviolet light on the skin. Popp was struck by the fact that the most effective wavelength for photorepair is 380 nm. He reasoned that the two phenomena, cancer being caused by chemicals that absorb light at 380 nm and the skin being healed by 380 nm light, could not be a coincidence. He concluded that there must be light in the body. With a graduate student, he built a detector using the then new charge coupled devices (CCDs). CCDs are the electronic chips now used in digital cameras and cameras in cell phones. At the time they were very expensive and were only being used in astronomical telescopes. Today they are everywhere, from cell phones to web cams.

Popp put plant seedlings in a quartz sample chamber and collected the light for a long time. He found that, indeed, living things emit light at 380 nm, Figure 2.

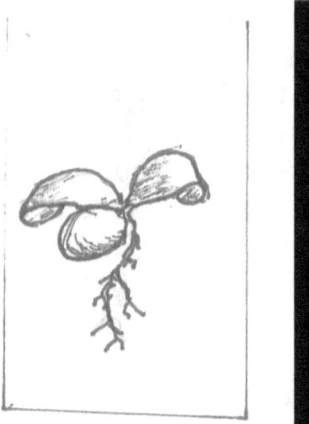 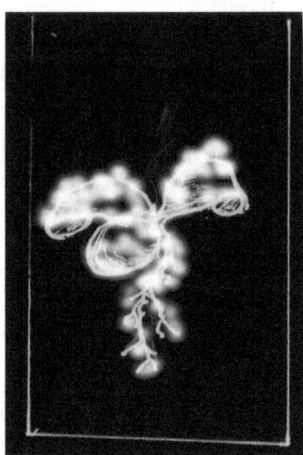

Figure 2 Biophotons from a plant seedling

He also found that the light is coherent. That is, all the photons being emitted from the seedling have the same frequency and phase, like light coming from a laser.

The suggestion that light plays a significant role in life processes is a serious, paradigm-changing idea. And, like most people speaking scientific heresies, Popp was treated very badly by his employer. He was asked to leave. Fortunately, Popp and his heresies have fared better than Gurwitsch. There are now many laboratories around the world working with biophotons and their interpretation.

Popp found that the biophotons are emitted by the DNA in the cells of all living things. Some of the photons escape from the organism as the initial photos of plant seedlings demonstrated. All living things emit biophotons. Primitive animals like worms and salamanders emit about 100 photons per square centimeter per second. Higher animals like humans only emit about 10 photons per square centimeter per second. This is a very weak light, like looking at a candle from four miles away.

Popp's biophotons account for Gurwitsch's rays. Gurwitsch's work tells us that not only do living things emit biophotons, they absorb them and are influenced by them as well. This was verified in 1972 with the work of S. P. Shchurin and his colleagues in Novosibirsk, Russia. They used tissue cultures in sealed containers separated by a glass partition. When they introduced a lethal virus into one of the cultures, only the cells in that culture died. When they replaced the glass partition with a quartz partition and again introduced the virus into one of the cultures, that culture died and then the other culture died. The difference between the two experiments is that quartz transmits ultraviolet wavelengths while regular glass does not. So biophotons are not just a status report of a living thing, they influence the life processes when they are received from outside.

Biophotons in people

Popp set up a fully dark room where all light sources were carefully excluded. In this environment, he could measure the biophotons being emitted by people and other large organisms. He found that living things emit light in the range of 200 to 800 nm wavelengths. This range includes all of visible light and some of the ultraviolet.

In healthy people, the light is coherent and varies with the many cycles present in living people. Those cycles include daily, monthly, seasonal and annual variations. He also examined unhealthy people and found that cancer produced a decrease in the coherence and order of the light. The rhythms were disrupted and the light was being emitted in different states rather than all being in the same coherent state. He found that multiple sclerosis was a disease of too much order. The light emissions were lacking a healthy variation.

Life as a quantum process

For our purposes the most interesting aspect of biophotons is the fact that they are emitted from all parts of the body in a quantum coherent state. This means that they are in this coherent state throughout the body. The fact that they are in the quantum state means that they are subject to quantum level influences and entanglements.

This quantum state is called superconductivity. We will look at it in more detail after we look at another kind of radiation in the body.

Biophotons tell us about the state of our health. If we interfere with them, with a chemical that absorbs them, for example, it causes cancer. It appears that biophotons are integral to the life process. Life, then, is a quantum process.

Communication between cells

Modern biology has found a long list of molecules called chemical messengers that are used to communicate within the body. The great majority of these molecules are proteins: hormones, neurotransmitters, clotting factors, histamines and antibodies. These substances regulate every aspect of our mind and body function.

The Emotions

From the first discovery of the messenger substances, the assumed mechanism of action was physical and chemical. Cells in one part of the body produce the molecules, they float around in the blood or in the intercellular fluids, and they finally land on a cell in other parts of the body. The cell recognizes the message that it has received and responds to it. To make this work it was assumed that there were receptors on the surface of the cell for each specific kind of messenger molecule. Candace Pert was the first to actually find one of these receptors, the opium receptor.[4] She was looking for neurotransmitter receptors which were assumed to exist only in the brain, but she went on to discover that there are neurotransmitter receptors on every cell in the human body. She looked outside the human body and found that other living things have the same receptors on their cells. Even single-celled organisms have the same neurotransmitter receptors that we humans do.

Pert has characterized the neurotransmitters as molecules of emotion, which is the name of her book. The neurotransmitters mediate our emotions. People demonstrate this every day when they take legal and illegal substances that alter the operation of neurotransmitters. They produce every kind of emotional state known, from abject terror to pure ecstasy.

Neurotransmitters got their name because they were first discovered in the synapses of the brain and it was assumed that their function was limited to the brain. With the discovery of neurotransmitter receptors throughout the body, it is clear that emotions are not just mental phenomena. With the discovery of neurotransmitter receptors on the cells of all living things, including single-celled animals, it is clear that what we call emotion and like to think of as a high-level human function is, in fact, a fundamental aspect of all life. This is a very important bit of information for our journey in this book. I will say in the next chapter that the medium for our connection and influence through the field is our feelings and emotions. This is our first hint that the patriarchy may have got it wrong about the value of feelings and emotions.

The radio signals

It is hard to understand how feelings and emotions can connect to influence things beyond our body if the emotions are carried by physical molecules and their effects in the body are created by a molecule physically sitting down in a receptor on cell.

Jacques Benveniste, working in Paris, provided an explanation. Benveniste was a researcher in allergy and immune response.[5] In 1984, one of his research assistants, Elizabeth Davenas, reported that she had made an error in setting up an experiment and had obtained an unusual result. The experiment was to measure the effect of anti-immunoglobulin E. antibodies (the reagent) on basophils. Basophils are white blood cells that have immunoglobulin E. antibodies on them. The error was to not put nearly enough of the reagent being tested in the solution. The unusual result was that the outcome of the experiment was the same as if she had put in enough of the reagent. This was puzzling enough to Benveniste that he began exploring how dilution of the reagents affects the results. He found that he could get the effect of the reagent even when the reagent was extremely dilute.

He found that initially as the concentration of the reagent was reduced, the impact of the reagent also reduced. But at a certain very low concentration he began to see an increase in the effects, even as the concentrations were reduced further, that is as there was less and less of the reagent chemical present. Another of Benveniste's coworkers mentioned that this looked very much like the results obtained with homeopathic remedies. Benveniste was not familiar with homeopathy, but he quickly found that homeopathic remedies at higher dilutions were considered more powerful therapy.

The challenge of these results was that it did not appear necessary to have any molecules of the reagent present in order to produce the effect of the reagent. This is definitely outside paradigm of someone steeped in modern molecular biology. Much to his credit (at least in the eyes of those rooting for the new paradigm) Benveniste went looking for what could cause this kind of result. From Benveniste's point of view, it cost him his career.

Benveniste eventually found two mechanisms that accounted for those strange observations. The first was that the interaction between bioactive molecules (like neurotransmitters and histamines) and cells is not carried out by a physical connection of a molecule landing on a receptor on the surface of the cell. The interaction is instead one of broadcast and receiver (Figure 3).

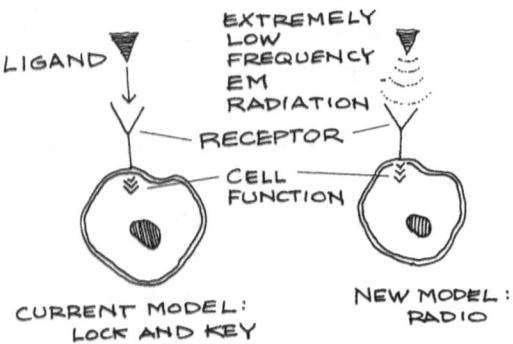

Figure 3 Interaction of messenger chemicals and cells

Bioactive molecules emit extremely low frequency electromagnetic radiation and cells receive and interpret those signals.

The second mechanism is that water can store low frequency radiation, so water can be used to carry the radiation between living things. That is a major deviation from the mainstream paradigm of biology so let's take a closer look.

Extremely low frequency electromagnetic radiation

What are extremely low frequency electromagnetic waves? Electromagnetic radiation consists of waves that come in a very wide range of frequency and wavelength. Visible light is the form we are most familiar with.

Chapter 9   Life is Light

Visible light is a rather narrow range of electromagnetic wavelengths. We call wavelengths a little shorter than visible light "ultraviolet" and this is what gives us sunburn and makes plants grow. Wavelengths that are a little longer than visible light we call infrared, which we feel as heat. We know other wavelengths of electromagnetic radiation that are longer than visible light as microwaves, television signals and radio signals. X-rays and cosmic rays are electromagnetic radiation with very short wavelengths.

Obviously, the properties of the electromagnetic waves vary a lot depending on their wavelength. Even very low-frequency waves are interesting. The United States and Russia use signals of 76 and 82 cycles per second, respectively, to communicate with submarines. The technical term for "cycles per second" is hertz (Hz). Waves at these frequencies have wavelengths of over 2000 miles. They have the very desirable property of being able to pass through earth and water. Here in the United States, the Navy broadcasts these low frequency signals from antennas buried in the ground in Wisconsin and Michigan. The waves pass through the earth and can be received by submarines anywhere in the world.

Benveniste found that the wavelengths being emitted by bioactive molecules were in the range of 20 to 20,000 Hz. This happens to be the same frequency range that we can hear as sound waves, which means that he was able to use conventional audio recording and playback technology in his experiments; that is, he used the sound card on his computer.

I should mention that sound waves are not the same as electromagnetic waves. A microphone converts sound waves into electrical signals, which is what recording equipment uses. An antenna converts electromagnetic waves into the same kind electrical signals that a microphone produces. If the frequencies are in the right range, audio recording equipment can record signals from either source.

The experiment that Benveniste ran thousands of times is this: He makes a solution of the chemical (an antibody, a clotting factor, a histamine) and places some of the solution in a test tube, which is then placed in a shielded coil. The coil is connected to an amplifier and then to a recording device. The signal emitted by the solution is picked up in the coil, amplified and recorded. Next, a test tube filled with pure water is placed in another shielded coil. The signal that was recorded from the solution of the reagent is then played into the coil around the pure water for 15 minutes. When that irradiated water is placed in tissue culture, added to a blood sample or injected into a guinea pig it has the same effect as if the actual chemical had been injected. These results have been replicated in other laboratories.[6]

To ensure that it was not some other local effect, the signal was recorded in one location and shipped via e-mail or CD to another, where the signal was played into a test tube of plain water and the water applied to the sample. The results were the same: The sample behaved as if the actual chemical had been applied.

Benveniste proposed that the mechanism of communication between a bioactive molecule and a cell is a form of the entanglement that we talked about when we were looking at the quantum weirdness experiments. If true, this means that like Popp's biophotons,

Chapter 9   Life is Light

intercellular communication in living things is a quantum level process.

Benveniste's work represents a serious challenge to the current physical/chemical paradigms of biology. The result was quite predictable. His work was vilified at the time and has been largely, but not completely, ignored ever since.

Connecting the mind-body

It is widely known, but not quite mainstream dogma, that our mental and emotional state influences our physical health. Descartes would be disappointed. Continuous stress causes all kinds of physical problems. We also know there are nerve connections to all the organs of the immune system, so that what we think can be transmitted to the immune system. This light in the body introduces a new mechanism providing communication between the mind and our health.

The microtubes again

The new connection medium between mind and body is that this light coursing through our body is carried by the microtubes, the same bodies that we saw earlier that provide the basic mechanism of consciousness. To consider what that means we need to examine another quantum process, superconductivity.

Superconductivity

Superconductivity is the phenomenon where, under certain conditions, electricity flows through a conductor with no resistance at all. Those conditions include rather exotic materials for the conductor and very low temperatures. Let us begin by considering regular conductivity.

If you want to make a magnet of the sort used in junkyards to pick up old cars, you wrap wire around iron cores. When you run an electric current through the wire, the iron becomes a magnet and you can pick up a junk car. When you turn off the current, the magnetism goes away and the car is dropped. It takes energy to make the magnet, so you have to provide a continuous current of electricity. Under normal conditions there is always resistance to the flow of electricity. You can feel the effect of that resistance if you put your hand on the cord to a hair dryer that has been running for a little bit. It will feel warm because there is a lot of current flowing through the wire. That heat represents a loss of electrical energy that is not available for drying your hair.

A simple description of superconductivity is that electricity flows through wires with no resistance at all. This means it will flow forever. It also means that if you have a piece of superconducting wire with current flowing and you can connect the ends of the wire to each other the current will continue to flow around the loop. Flowing current is what makes a magnet. The application of superconductivity that most people are familiar with is the MRI machine in hospitals. Making an image with magnetic resonance requires very strong magnets. Superconductors can produce stronger magnets than regular wires so that is what is used in the MRI machines. The reason the machines are so big is that the superconductors work at the temperature of liquid helium, which is 4.7 C above absolute zero. A lot of insulation is required to keep the liquid helium from

Chapter 9 — Life is Light

boiling away too fast. It is the cryogenic insulation that makes the MRI machines so big.

Once the current starts flowing or superconducting, no further current input is required to keep the magnet operating. It is cheaper to buy the liquid helium than to buy the electricity it would require to produce those magnetic fields with conventional electromagnets.

Superconductivity is a fascinating topic. It seems to have generated a very large number of Nobel prizes throughout its history. But it is not the flowing electron model of superconductivity that interests us here. The idea of electrons flowing without resistance is a nice model for superconductivity, but it is not entirely accurate. Superconductivity is actually a large-scale quantum effect. The electrons are not actually flowing. Rather they are everywhere they can possibly be all once: i.e., they all share exactly the same quantum state. It is like they are all the same electron. We call it a large-scale effect because this single electron is spread over the entire length of the superconductor. Superconductors are used in the rails of maglev trains (trains that float on magnetic fields instead of wheels), so the size of the quantum "object" can be as long as the conductor in the rail.

In this quantum state all of the electrons are entangled and distributed. Anything that affects one part of the quantum object, the collection of electrons, affects all parts instantly. I'm going to suggest that the light in our bodies that represents the state of our health is in a single, coherent quantum state.

Super radiance

"But still," you might say, "superconductivity is just electrons, and life is a much higher level phenomena than just electrons."

That's certainly true. The photons that we have been talking about in biophotons and Benveniste's extremely low frequency radiation are also well below the level of life. But we are not concerned with the physical electrons and photons themselves. Our interest in them is based on the information they carry. Photons and electrons are waves and waves can carry information, for example our radio and television signals and our cell phone conversations. We saw that holograms recorded from light waves carry a tremendous amount of information. The holograms in our brains, made by oscillating electrical charges, also carry huge amounts of information.

So a collection of coherent, superconducting electrons or photons can serve as a pool of information. The name for this kind of light in the body is super radiance. Let's consider how that works in the human body/mind.

The body/mind

We have said that light at various wavelengths is produced in the body. Benveniste said that light gets conducted or transmitted from one place in the body to another. How does that work? If you turn on a flashlight and put your hand over the light and then look at the back of your hand, you can see a little bit of light coming through your hand, but not a great deal. Based on that observation, it appears that the body is not a very good conductor of light.

Chapter 9 — Life is Light

The folks at the Center for Consciousness Studies at the University of Arizona have proposed that our old friends the microtubes are excellent conductors of light. The reason they are good conductors of light is not that they have shiny insides. Instead, when a photon is moving through a microtube it is maintained in the distributed, quantum state. Which means that the photon is not flowing through the microtubes, it is simply everywhere it can possibly be, all at the same time. In other words, microtubes are superconductors of light.

Everywhere it can possibly be, in this case, is our entire body. Popp observed that the biophotons being emitted from the body are all coherent. This means that they must be coherent inside the body as well. So all of the photons in the body are coherent and in the quantum distributed state. This is how we described superconductivity in magnets. This tells us that our state of health, our entire state of being, is carried in a single macro-scale quantum object that is at least as big as our physical body.

Chi and biophotons

If this idea, that our state of being is carried in light flowing through the body, seems strange to you, I should point out that it supports a very old idea. The old idea is acupuncture and our chi or life force flowing through the meridians. The Taoists proposed a long time ago that our state of health is represented by chi flowing through channels in our body called meridians. Where the meridians come close to the skin, we can access the meridians with needles, pressure or magnets. These are the acupuncture points. Health is the free and even flow of energy through the meridians. Disease is a blockage or imbalanced flow of energy.

The normal view of microtubes is that they are responsible for motion and structure of cells. There are microtubes in every cell in our body. I have mentioned that they are common in neurons because the neurons can move. There are also lots of microtubes in the structures of our body that give us form, such as our skin. So it is not entirely surprising that the Taoists were able to find an abundance of life energy near the skin. Further experimental support for this idea of light being conducted in the body comes from experiments where bright light has been shown on one acupuncture point and a light can be observed emerging from another acupuncture point on the same meridian.[7] The work was done at the Institute for Clinical and Experimental Medicine in Novosibirsk, Russia.

Here's another example where modern Western science has caught up to Eastern practices that are thousands of years old, although I doubt that many practitioners of Western medicine would consider research from Russia part of Western medicine.

Light in life

It appears that this life light is not just a passive repository of information. This single coherent quantum entity appears to be involved in communication within the body. Experiments measuring the time required for the exchange of information between the brain and the extremities yielded some interesting results. It was found that the time between stimulus in the brain and the arrival of nerve signal at the hand was consistent with the signal being transmitted through the nerves. But when the time was measured between applying a stimulus on the hand and the response appearing in the brain, it was found to be zero. The brain responded instantly to a stimulus applied to the hand. This is very difficult to account for if the only communication mechanism is nerves, which have a finite transmission speed. It is much easier to understand if we are dealing with a coherent quantum object in the body. In this case any change made to any point in the light body is instantly conveyed to all other points of light body. The reason is not that the transmission speed is infinite, but rather it is because all of the components of the light body are the same component.

Growth, form and light

Here's another mystery that super radiance can solve: development and form in living things. The conventional wisdom today is that the DNA in cells directs the development of the organism, whether a single cell or an adult human with millions of specialized forms and functions. Most living things show great symmetry. Our arms and legs are about the same length, our feet and hands are approximately the same size, not exactly the same size, but very close considering how far apart our hands are on a cellular scale. There must be a great many very sensitive feedback loops operating across very large distances for organisms to develop from a single cell to the functioning adult with such large-scale control over form and symmetry. Molecular biology offers no suggestions about the mechanism of this amazing feedback and control system.

But if our cells are communicating via coherent light, this provides at least a medium for the kind of large-scale control and feedback that obviously goes on all the time. This may provide an explanation for some of Sheldrake's morphic fields.

Connecting light and consciousness

Thus far the light I have described is involved in low-level communication, that is, signals between chemicals and cells and between cells and other cells. It seems to me that there is the possibility of the light being used for higher-level communication, between our human-level thoughts and feelings and the rest of our bodies. Nature doesn't do things for frivolous reasons, so it is unlikely that it is a coincidence that the microtubes serve both as a mechanism of consciousness and intelligence on one hand, and the medium that conducts the life light through our body. If there is any interaction between the microtubes' function as a center of consciousness and intelligence and their function of conducting the light body, then we have a very intimate connection between our thoughts and feelings, on one hand, and our complete state of being including our physical health on the other.

This connection represents the second personal speculation that I introduce in this book. The first was the connection between the macro-scale holograms of the mind and the quantum field via the microtubes. I have no technical basis for making either conjecture. The reason I introduce them is that with these conjectures I can account for a great many observations. According to the rules of science, that is an entirely sufficient justification for making those conjectures.

Life is light

We see from the work of Popp, Benveniste and others that life is conducted with electromagnetic radiation or, to put it a little more poetically, life is light. This light is produced and consumed within the body to regulate its function. The light is also broadcast out of the body and can be received by other beings, where it influences their state. The light is coherent throughout the body when the body is healthy. Coherence means that all of the light is in the same quantum state. What this means here is that life and all of its functions are quantum level processes. This all is in addition to the quantum nature of our thinking and feeling processes. So everything about our lives is subject to the same quantum level knowing and influences.

Light that is coherent throughout the body is a "body level" phenomena. The function and purposes of the light cannot be seen by looking at the individual components, or chemicals of the body, which is the standard practice in the life sciences today. We can say that life and the light are emergent properties of the whole system. When you take the system apart, the emergent properties are destroyed. I think that emergent properties like the coherent biophotons would provide an excellent basis for a definition of life, but adopting such a definition will have to wait until the mainstream paradigms catch up with the current research.

We will see later that what people call spirit and spiritual is exactly the quantum field side of our existence. We have seen in this chapter that our bodies also exist in the quantum field. I should be more precise: the living information part of our bodies exists in the field, that is, or health, body sensations and feelings. This makes our bodies as much a part of the spiritual world as our minds. The spiritual practices that include and honor the body are closer to the truth.

The next steps

Our thoughts and feelings and our physical health are all waves in the quantum field. A couple of chapters ago we said that those same waves determine the probability curves for the outcomes of events in the material world. I can combine those two ideas and say that thought/feeling is action at the quantum level. There are more popular ways to describe that: influence, mind over matter and action at a distance. The model we have developed here supports the reality of influence at a distance. But before you get out your Harry Potter robes and wand and start turning your enemies into frogs, you should know that the model supports a very limited form of influence, particularly compared to Harry Potter's style of influence. We take up influence next.

How far have we come?

Last time we checked our progress we had made the mind into a quantum process. Now we have extended the quantum process to include the whole body. We found that light in various forms maintains our health and tells us about our state of health. We also broadcast the light out to our environment.

Communication between cells is also carried by electromagnetic waves, this time by very low-frequency waves. Our whole body runs on light.

All this light is carried in the microtubes and is super radiant. It is all in a single quantum state. Our state of being is carried in a single patch of quantum coherent light. The connection of the light to the microtubes provides a way for our quantum intelligence to control our physical body. The loop is closed. There is no separation between our transcendent souls and our material bodies. Science and the church are gong to have to work out a new deal to replace Descartes'.

---

[1] Gurwitsch, A.G. *Mitogennetic Analysis of the Excitation of the Nervous System*. N.V. Noord-hollandsche Uitgeversmaatschappij, Amsterdam, 1937.

[2] Tompkins, P, Bird, C *The Secret Life of Plants*, Harper-Collins, New York 1973 page 55.

[3] Popp, F.A. "Biophotonics: a powerful tool for investigating and understanding life" in Durr, Popp, Schommers (eds), *What is Life?*, World Scientific Press, Singapore, 2002.

[4] Pert, Candace, *Molecules of Emotion*, Simon & Schuster, New York, 1999.

[5] Benveniste, J., "From water memory to digital biology", *Network: The Scientific and Medical Network Review*, 1999; 69: 11-14.

[6] Sainte-Laudy, J., Belon, P. "Analysis of immunosuppressive activity of serial dilutions of histamines on human basophil activation by flow symmetry", *Inflammation Research*, 1996, Suppl 1:S33-4.

[7] Pankratov, S. "Meridians Conduct Light", *Raum und Zeit,* Germany, 1991.

Chapter 10
Influence in the Quantum Field

B Y INFLUENCE, I mean affecting the outcome of events with no physical contact. Action at a distance and mind over matter are other ways to put it.

It is a tricky topic. On one hand, it is a very desirable thing to do if you can do it. Witness the popularity of *The Secret*, prosperity groups and Harry Potter. On the other hand, it is dismissed and condemned with great vehemence by our mainstream science and by the culture at large. Part of the reason for that condemnation may be that action at a distance is the province of gods. And in science, anything related to god is not allowed.

But quantum physics has demonstrated that Einstein's "spooky action at a distance" is a real, observable quantum level phenomena. The last couple chapters have demonstrated that life is a quantum level phenomena, so living things should be able to do action at a distance. And so they can. And so they do. Let's look at how that works. I'll begin with the nature of influence at the quantum level and then take up how we humans exert that influence.

Action and influence in the quantum field

In Chapter 5, I said that the quantum field is a hologram of the space and time in our normal material world. In Chapter 7, I said that our thoughts, feelings, emotions and memories are carried as holograms in our brains. Then I proposed that microtubes in our neurons translate between the macro-scale holograms in our brains and the micro-scale hologram of the quantum field. This mechanism accounts for the non-material phenomena of knowing at a distance and influence or action at a distance. My suggestion that the quantum field and quantum level behaviors are involved imposes some significant restrictions on how the influence works.

Power!

I think it's important to discuss these restrictions because of the very widespread and long-standing desire among humans to have great personal power or at least to have spiritual allies who have great power. The myths and legends from the ancient past are not all that different from the current action hero movies. Proponents of many religions, in the past and in the present, describe their deity as being very powerful in the material world. It is also popular to imply that members of that religion can acquire some of that power through their membership. Material power remains very popular. The promise of gaining material power remains attractive in spite of the fact that long experience shows that no individual or organization seems to be able exert much non-material influence without doing a lot of very material work.

Chapter 10 — Influence in the Quantum Field

In contrast to the very popular cults of personal power are the traditions that advocate nonattachment to outcome and being in the flow. These are mostly mystic traditions and I believe Taoism is probably the best example. These traditions are not nearly as popular as the traditions that trade in material power.

I mention this because the implication of using quantum level "power" for nonmaterial communication and influence is that the influence is subtle. So how does it work?

Quantum Level action

The nature of action at the quantum level is very different from action at the level where you and I live. The table below compares the two domains.

Material World Behavior	Quantum, frequency domain behavior
Massive objects, predictable outcomes	Low to no mass objects, probabilistic outcomes
Everything is local and temporal	Nothing is local or temporal
Cause and effect: no effect without a cause	Individual outcomes are probabilistic
Physical force needed to change an outcome	Observation alone can change outcomes
Objects are isolated until acted on by a force	Objects are entangled and affect one another over all time and any distance
Only one outcome for a specific initial condition	All possible outcomes for any initial condition

Table 1 Macro-scale and quantum scale action

Life and Spirit in the Quantum Field

In our normal walking-around world, things work pretty much the way Newton said they do. Nothing changes unless it's acted on by some force. Things with mass, like you, me, squirrels and planets, are separated from one another by time and space and they can only influence one another through physical contact or one of the two physical forces that operate at that scale, gravity and electromagnetism. When we apply force to something, we can pretty accurately predict what the outcome will be.

Moving the curve

The quantum equations tell us that quantum-level forces only work at the level of probability distributions. That is, when you calculate what is going to happen in a given situation, the answer you get is not a definite result, it is a probability curve. The probability curve defines what range of outcomes are possible and the percentage of each type of result. This tells me that at the quantum level, "forces" have the effect of moving the probability curve, rather than changing a specific outcome like they do at the level of large-scale material things. The influence that is transmitted through the quantum field can only alter the probability distribution of the outcomes.

That last sentence is a little abstract. What does it mean in concrete terms? Quantum-level influence can, at best, move the probability curve for the behavior or health of individuals, for example. That means that all of the human behavior that is possible, from saintly to abject evil, will still happen. What might change is the percentage of different types of behavior that we see, but there will still be good and evil behavior.

Chapter 10   Influence in the Quantum Field

For those who find that assertion bothersome, you don't have to look very far to see that human behavior always spans the complete range of what is possible: from good to evil, callous to considerate, enlightened to paranoid.

Nature of quantum level action

There are two things to remember about this discussion of quantum-level action. One is that quantum-level influence is probabilistic. If we are just running the experiment once, we cannot predict what the outcome will be, even if there is significant quantum level influence directed toward a specific outcome. The situation was nicely illustrated by the BBs and the pegboard example in Chapter 5. If I pour lots of BBs down through the pegboard, they will come out in the normal probability curve. I can predict the shape of the curve, but it takes many individual events, that is BBs, to make the probability curve visible. If I only drop one BB down the pegboard, I can say nothing at all about where it will come out.

The other important point is that quantum-level influences can only (or usually) affect quantum-level events. Quantum-level events are things like the movement of isolated photons, electrons, protons and neutrons, and random events that are determined by extremely small variations in the physical arrangement, like the BBs bouncing down through the pegboard. What makes this interesting to us is the fact that life and all of the life processes are also quantum-level events so they are subject to influence from quantum-level forces.

Humans and other living things can exert influence through the field because the probability waves that

Life and Spirit in the Quantum Field

influence outcomes are the very same waves that carry our thought/feeling. The rational guy in me likes to say, thought is action in the quantum field, but that is not quite the way it is, as we will see. It is more accurate to say that feeling is action in the quantum field. Let's look at that idea.

Our connection to influence

All this quantum science has been very interesting to physicists over the last hundred years. The similarities between quantum level reality and spirituality have been noted by many people, but until now there has not been a "material" connection between macro-scale life and quantum-scale phenomena. The mechanism I described here, the microtubes connecting our macro-scale holograms to the micro-scale hologram of the quantum field, provides that material connection. At the quantum level, thought is the same as action. Our thoughts, feelings, emotions and intentions are quantum-level objects. They can therefore be distributed across time and space and interact with other quantum objects.

The important point in all this is that our thoughts and feelings can initiate quantum level action. Quantum level action includes human health, behavior and decisions. This connection between living things and the quantum field is the mechanism that underlies all of the nonmaterial phenomena that are such an important part of the human experience.

If this is true, it imposes a serious limit on the kind of "force" that the non-material events can exert in our normal, material world. Later, I will propose that god is a quantum-level intelligence, which means that god can exert "only" quantum-level influence. I hope it is obvious by now that that is a very profound influence. Profound or not, I expect this may be upsetting to those who are hoping that their God is all-powerful in the material world. For my part, I find it encouraging in that it accounts for subtlety of the nonmaterial phenomena that has been observed for as long as we have records.

There is an interesting paradox here in that the objects that behave in this strange quantum way (individual photons, electrons, neutrons and protons) are the building blocks of all of our normal, massive, Newtonian reality. When those quantum objects start clumping together, they start to lose their quantum behavior. But as we have seen, when the clumping reaches the level of living things, quantum behavior again becomes active.

The engine of influence: thought and feeling

I have been using the word "thought" a lot when talking about influence. That reflects my Western, guy upbringing. I am hardly unique in this bias toward rational thought. But it turns out that it's the feelings that are important, not the conscious, logical thought. Many readers, mostly women, will think that last sentence is perfectly obvious. Others, mostly guys, may think it strange. Perhaps because I'm one of those guys, I would like to present some logical reasons for why feelings are important. Is that a contradiction? Nonetheless, here are my reasons.

Feeling and behavior

Let's begin with regular actions and behavior in our everyday lives. We like to think that we are conscious beings who decide what to do and then do it. In other words, our behavior is guided by our conscious thought. Remember the patriarchy. Even if we let ourselves go and give in to our feelings and emotions once in a while, we like to think that we really run our lives based on our rational thoughts and decisions.

While I am sure that is true for the present company, I'll bet most of you know of people who have trouble and pain in their lives because of the bad decisions they repeatedly make. When you talk to those people, many of them can tell you what they know they should be doing, but for some reason don't do. Logically, they know what to do, but they continually make other decisions that seem to lead to trouble and pain.

As I have been writing these pages, there have been regular news stories about men in positions of great power and responsibility in government and churches. These are rational, reasonable MEN, who are usually married to attractive women, but for some clearly irrational reasons have chosen to go off and have affairs with other women, often at government or church expense and often after loudly condemning other men who have done the same thing. Where is the reasoned, logical, rational behavior that is the hallmark of real men? The answer is that behavior based on logic and reason is a myth. Everybody's behavior is driven by feelings.

Why do some people seem to be models of consistent, reasonable, rational behavior? I think the reason is that in those people, their inner feelings are aligned with their logic-derived needs and wants. It's not that they have found a way to let their logic triumph over their feelings.

Chapter 10 — Influence in the Quantum Field

The importance of sincerity

People have recognized that it is possible to influence events in the world. This is the basis for prayer in most all of the world's religions. Of course the influence is not a 100%-Harry Potter kind of influence (more on that in the next section), but there is enough truth behind the belief to keep it alive for all of recorded human history up through the present.

There is the idea among advocates of the effectiveness of prayer that the prayer has to be "sincere" or "heart felt" in order to be successful. Idle wishes don't work very well. This observation supports my suggestion that logical thought is not the source of the influence.

This old truism about prayer is nicely supported by another group of people who would exert influence on the world through their inner efforts. These are the people in prosperity groups and fans of *The Secret* in its book and video forms. For the uninitiated, prosperity groups are populated with people who are trying to increase (usually) their material prosperity by thinking positive thoughts. There are local meetings, webcasts, workshops and conferences. I don't know how many people are involved, but it appears to be a large number.

The Secret book and DVD can be considered as user manuals for prosperity groups. They have been very popular, rising to #1 on the NY Times best seller list and #1 at Amazon, respectively. The emphasis in *The Secret* is that conscious thought is the basis for exerting influence in the universe. Most of that influence is aimed at acquiring money, large houses and trophy spouses.

I have watched the DVD and I can report that, to their credit, they do say that feelings really do need to be aligned with the outer goals and that many of the feelings that guide people are subconscious. But those were just single, passing references. The great bulk of *The Secret* is devoted to the idea that conscious thought is the basis for our influence in the world: Just think positive thoughts for a few days and –bling!–you're rich! This is the reason for its great popularity. Our consciousness-oriented, patriarchal culture does not like to contemplate inner feelings of any sort.

So what evidence does the example of *The Secret* offer to support my assertion that feelings are the drivers of influence and not thoughts? For all the millions of people who have eagerly read *The Secret* and participated in prosperity groups, I have not seen any big uptick in the number of newly wealthy people. There has been no run on all the McMansions available on the market. The wedding announcements seem to be full of the same regular people with no big increase in trophy spouses. People wish to influence their condition and their future for the better, but they also want to do it with conscious thought, which is why *The Secret* is popular. It also explains why it has not been successful in producing the number of transformations hoped for by its fans.

Examples from healing

Another line of evidence supporting feelings as our means of influence comes from the energy healing community. Energy healing is a form of influence (over a person's health) at a distance. The widespread advice given during classes on energy healing is that your inner state is the key to your success as a healer: specifically, a centered and relaxed state is essential for successful healing. This state is more of a feeling than a thought, or more accurately, it must be felt to be real. I have had training in two kinds of energy healing, Therapeutic Touch and Healing Touch. I have been around a variety of healers, from gifted to awful. The awful healers are always head people. They are not feeling the healing intent in their bodies.

The difference between the two kinds of healers is immediately obvious to the person being healed. When someone who is not centered and relaxed is working on you, it feels jarring or irritating. When I was taking Therapeutic Touch classes, I was usually the only guy in the class. Being an old engineer and starting late in this intuition business, I was far behind most of the women in the class, many of whom were experienced intuitives. At one point we paired off and were to do an exercise where we would make three passes with our hands over the backs of our partners. I did the first two passes and realized that I was anxious. The anxiety generated thoughts like, "Can I do this?" and "Will I feel anything?"

I recognized that I was not centered and so took a moment to center myself before doing the third pass. Centering is a feeling, one of relaxing and softening in the body. I then did the third pass. At the end of the exercise, my partner said that the first two passes felt pretty awful, but that the third pass felt very good.

Being "in our heads" does transmit something across space, but it is not good. The feelings that underlie conscious questioning, doubts, anxieties and such do not feel good. Transmitting good things across space requires being in our bodies, and specifically, being comfortable and relaxed in our bodies.

I have used the persistent old wisdom before in this book as an indicator of underlying truths. One of the most basic truths that spans all religions and cultures or perhaps more accurately has been taught by the founders of all religions, is the great value of forgiveness, acceptance, compassion and love. Those are all feelings and emotions. They are not thoughts. They seem to me to be four degrees of the same underlying feeling. There is universal recognition that they must be sincere, felt in the body and heart-felt to be powerful. The old wisdom is that these four feelings are subtle but profound forces in our world.

For all of these mostly logical reasons, I conclude that feelings really do make the world go round. Feelings drive our conscious behavior in the material world. They provide our means of connection to information in the field. Most importantly for this section, they are the basis for our influence in the quantum world.

If it is feelings that connect us to the field, then we have to say that the field feels. The fabric of reality, the ground of all being is feelings. Think about that, guys.

Probability curves and the limitations on influence

So now we're in the right inner state. We can move mountains, right wrongs and raise the dead. Right? Well, maybe. When we talk about exerting influence, we are usually thinking about some specific situation and some specific outcome that we want: recovery of our son or getting that job or turning the boss into a toad. It is precisely here that the limitations imposed by the quantum model come into play.

Chapter 10 Influence in the Quantum Field

The influence that we, or anything else, can exert at a distance is a quantum level force that acts on quantum level events. Events in the large-scale material world, the world we walk around in and bump into, mostly obey Newton's laws of action and reaction exerted by the regular physical forces. They are extremely difficult to influence with quantum level forces. Fortunately for us, there are lots of quantum-level events available to influence, like choices (and thus behavior) of humans and other living things and the health of living things.

These quantum level events in living things are guided by the probability curves that make up the quantum field. In conventional quantum mechanics, the waves of the quantum field define the probability curves for the outcomes of events we see in the material world. For massive things, the probability curve is very narrow and produces the behaviors that Newton predicted. For things with little or no mass, the curve is much wider, so that everything that can happen does happen. In the case of outcomes of events that involve quantum events, like human behavior, all possible outcomes and behaviors are possible. Our influence cannot "force" a specific outcome. The outcome of a specific situation is always probabilistic.

What does "probabilistic outcomes" mean for someone who is praying for a better job or for a healer who is trying to prevent the death of her patient? It means that even with the best skills, the greatest centering, the most sincere intent, the outcome in any specific situation can be anything, from not getting the job to getting the job with a bonus, from the patient recovering to the patient dying, and everything in between. It is difficult to tell what will happen before it happens.

Life and Spirit in the Quantum Field

The event I would like to consider is a healing event. A healer is working on a patient with a serious problem. The possible outcomes for the event include: dies, lingers for a time and dies, recovers with a disability, recovers slowly, recovers quickly and fully. That covers all the possible outcomes. Imagine that we can calculate the probability curve for the outcome of this event, before the arrival of the healer. It is shown in Figure 1.

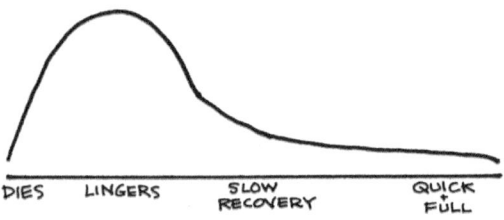

Figure 1 Health event outcomes curve

In our imaginary calculation, the shape of the curve includes many influences, including the patient's own feelings and expectations, his or her family, friends, community and anyone else with an interest in that person or even that class of person. The particular curve I drew favors a less-favorable outcome, but all of the outcomes are possible. A probabilistic outcome means that for this specific person, we don't know what the outcome will be. If we looked at many people with the same sort of outcome probability curve, most would linger or die and a few would recover.

Now the healer intervenes with well-centered, sincere and detached intent, which is a wave of the same medium as the probability curve. Because she is directing her intent toward that particular patient, her intent connects with the probability curve for the patient

Chapter 10   Influence in the Quantum Field

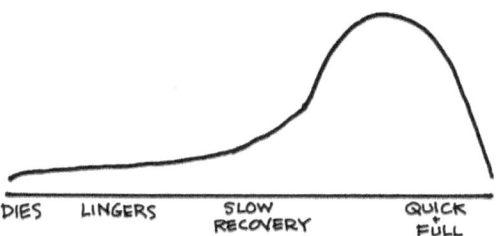

Figure 2 Outcomes with a healer's intervention

and interferes with it. The result is an altered probability curve for the patient's outcomes, shown in Figure 2.

The altered curve looks more favorable for the patient, but all the original outcomes are still possible. The change in the curve from Figure 1 to Figure 2 is quite dramatic as healers go. That kind of change in probabilities would be the result of a very powerful healer. Even with such a healer, all possible outcomes are still possible. What will happen with this particular patient? We can't tell until it happens. That is the essence of what "probabilistic outcome" means.

Scientific Approaches

An aside on observing trends in distributed outcomes: the only way to see, in the material sense, the effects of the prayer or the healing intent is to look at the outcomes over many experiences. This will fill in the probability curve for the event. The science people among the readers of this book will say, "Of course!"

Much of the research into psychic phenomena has used many, many trials. In spite of the scientific merits, there are several problems with this approach. The involvement of the researchers, usually with their very intellectual, if not skeptical, inner states is a negative influence on the experiment by the very same mechanisms that the healer and prayer are invoking. Another problem is the sheer number of repetitions called for in conventional science experiments. It is not pleasant or encouraging for the participants. Conventional scientists tend to treat their human subjects like machines. So while I believe that in order to see the effects of intention in the material world many repetitions are required, I am not advocating the rather insensitive approach that is common in research today.

Unity of thought and action

A note on feelings: Even after my logical arguments earlier about why feelings and not conscious thought are the medium of the field, I find it is very hard for me to come right out and say that. I was comfortable writing the heading of this section as "Unity of thought and action", but I found I was uncomfortable trying to change it to "Unity of feeling and action." Saying "thought/feeling" as I do in the following paragraphs was OK, but taking out the "thought" part is definitely a challenge. I am a child of Descartes and the patriarchy. So rather than cause myself all sorts of discomfort, I will just ask you, dear reader, to substitute "feeling" and "feels" whenever I say things like, "the field thinks", or "thought is action" or "thought/feeling." I appreciate your help.

I have also argued that the waves of the quantum field are the medium of thought/feeling. It is the same medium that we use in our macro-scale minds. Since thought/feeling has no mass, it is distributed, which nicely accounts for all the non-local psychic phenomena we have been talking about. But it also means that "thought is action" is literally and materially true. The medium of thought/feeling is exactly the same thing as the medium of the probability curves that guide the outcome of events in the universe.

So our thought/feelings about a particular subject contribute directly to the probability curves for the outcomes related to that subject. I say contribute to because for most events more than one person is thinking/feeling about the event. All interested parties contribute to the shape of the probability curve, and then the actual outcome of any specific event can be anywhere under the curve.

Limits on outcomes

The practical result of this state of affairs is that things don't always come out the way we wish they would. On one hand, that is hardly a radical statement. It is everyone's normal, everyday experience. On the other hand, it imposes a profound limitation on the kinds of results we can expect from any form of action at a distance. Saying that the possible outcomes for an event are distributed under a probability curve means that sometimes you get the result you want and sometimes you don't. This is not the result of "not doing it right" or "not being good enough," it is just the way things are.

If everyone already knows this, what difference does it make to say it here? It is important to know when we listen to claims made by people selling action at a distance in any form. Who sells action at a distance? Some churches, fundamentalist evangelists in all religions, energy healers and energy healer trainers, people selling *The Secret, The Law of Attraction* and prosperity groups and people selling much of the self-help material. It is very common to say or imply that whatever is being sold will produce deterministic results, when at best they can produce probabilistic results. So are all these people being evil by selling something they can't deliver?

It is important that we all know, and know in our gut, that outcomes of events are probabilistic and that they won't and can't come out any particular way all the time, or even very much of the time. The reality is that we "know" that in our intellects, but most of us harbor subconscious desires for Harry Potter-esque, deterministic, miraculous, magical outcomes. Marketers of all stripes know that about us and we happily buy their products when they promise, or even subtly imply, magical results. On the other hand, I doubt that anyone would buy cars or shampoo promoted by ads like, "You have a 0.2% chance of improving your love life when driving our Model X-40 convertible."

In the end it is possible for an "influencer," like our job seeker or the healer, to alter the shape of the probability curve for a situation, so we can say that it is possible to influence the outcomes of situations. However, it is important to understand that that there are no isolated events. Every event is available to anyone who has any interest at all in it and all those people make some contribution to the shape of the probability curve. We have the vagaries of a probabilistic outcome added to the hordes of people who can contribute their influence to the shape of the probability curve.

Action at a distance

Non-material influence of the material world is real. It works through the quantum field that is both the medium of feelings and the medium of probability waves for quantum level events. The result is that the influence is much more subtle than people trying to sell power would like to admit and much more profound than the advocates of Newtonian materialism would allow. In the middle are all the regular people who have always recognized the reality of action at a distance and have used it in all aspects of their lives.

After all this we are back where the mystics started. Large-scale material power can only be exercised with large-scale, material forces. The influence at a distance that we, as individuals, have is subtle but profound. The path to that influence is through self-awareness, acceptance and non-attachment to outcome. The master swordsman carries no sword.

How far have we come?

We have talked about many mechanisms up to this point. In the chapters that follow, I'll be applying those mechanisms to how we live our lives, so I thought it might be helpful to review the mechanisms before we go on.

Physics of the quantum field

- The quantum field is the ground of our physical reality.

- It is a hologram of both time and space.

- The field is the origin of the movement and interaction we see in the material world.

- Objects with high mass, bigger than 60 carbon atoms, are local and temporal in the field.

- Objects with low mass are distributed across time and space in the field.
- The waves of the field define the probability curves for the outcomes of quantum level events.

The form of life, human thought/feeling and health

- Human thought, feeling and perception are carried as holograms in the mind.
- The form is very similar to the form of the quantum field.
- In the perception process, fragments of holograms from sensory inputs are used to retrieve holograms from sensory memory. The combined, retrieved fragments are projected back out the sensory channel to form the image/sound/taste, etc. that we perceive.
- Our state of health, and probably information about the form of our bodies, is carried by quantum coherent electromagnetic radiation in the body that is in a hologram, called super radiance.

Connection to the field

- The macro-scale holograms in the mind and body are connected to the quantum field by the quantum oscillations of the microtubes in all cells.
- Thought, feeling, memory and state of health are mass-less objects distributed across time and space in the quantum field hologram.
- Our thought originates in the quantum field, like the motion of particles originates in the field.

Chapter 10 Influence in the Quantum Field

- The same waves that define probability curves for particle behavior provide thought and memory for living things, including humans.

- The quantum field is holographic memory: all information is content addressable.

Because of the connection to the field

- We can access our memories in the field.

- We can access any information distributed in the field, limited only by our ability to interpret the information with our projection perception system.

- The field is thought, the field thinks/feels.

Influence through the field

- Thought is action at the quantum level because the same waves that define probability for outcomes of quantum level events are the waves of thought/feeling.

- Quantum level events include feelings, decisions and health of humans.

- The influence we exert is to move the probability curve for the outcomes of quantum level events, which is subtle but profound.

How does all this play out in our daily lives? Like action at a distance, the possibilities for change in our lives are subtle, but profound. Let's look.

Part 3: Quantum Life: The Possibilities

We began our journey back in Part 1 by describing a fairly liberal view of reality, one that included many capabilities that our current, Newtonian mainstream science and culture do not allow. We then applied the principles of quantum mechanics to life and found that we humans have some entirely amazing potentials. Things are possible that would have shocked Newton and the boys. Actually, that's probably not true given what we know about Newton's spiritual life. We can say that those capabilities would have shocked the Royal Society of Newton's time.

It turns out that many of the paranormal, alternative and spiritual practices that are part of the normal human experience are, indeed, possible with the mechanisms of quantum life. These days we have the luxury of being more inclusive since we don't have to deal with the Inquisition.

The new science tells us how the paranormal practices work. When we understand how these things work, we can learn how to use them effectively. We can also unravel truth from fiction in claims made about those practices. The truth is quite amazing, so you won't even miss the fictions.

This part focuses on things people can do. The next part will look at possibilities in the spiritual domain. I think the material and spiritual domains are really the same thing. There is only one domain, but we have a lot of history around these topics, so I will move slowly. We will take up spirit in the next part. For now, lets look a how human abilities work and how to use those abilities.

Chapter 11
Thinking in the Field

THE MECHANISMS I described in the last part imply a broad range of abilities that we humans have. If you are a confirmed Newtonian materialist, these will be wild new ideas. If you are a Reiki master, they will be the normal, regular things that you do.

The first ability I would like to consider is knowing, that is, receiving information from the field. Since the information is not received through the regular senses, it is called non-sensory knowing or intuitive knowing. That makes it sound like the field is a big library in the sky. It is full of information that we can look up. We will see that we can receive new information, which tells me that the field is much more dynamic than a library. It is a living, feeling, creative entity. That idea will have profound implications when we get to the part on spirituality.

Let's begin with the library idea.

Information

I described holographic memory as being "content addressable." That means that if I have a small fragment of the information I am looking for, I can "project" that fragment into the hologram and I will get back a bigger piece of information that contains, or somehow matches, the initial fragment. Human memory is often described as being based on associations. When you get a whiff of apple pie baking, your mind is flooded with memories of grandma's kitchen at Thanksgiving. If you want to remember the names of lots of new people, it helps to make some association with each name, like "Joe Smith has bushy eyebrows," or "Al Grant looks like humpty dumpty." Having the association makes it easier to recall a person's name.

"Association-based" memory is what you get when you have content-addressable information. The input fragment serves as one side of the association and the information that is returned is the other side. Human memory is associative precisely because it is holographic memory in the field. All human memory, except short term, which may be local holograms, is stored holographically in the field. That includes long-term memory and all of the sensory memories that we use in Pribram's perception process. The projection of the fragments and reception of the results is carried out by the quantum oscillations of the microtubes immersed in the physical holograms maintained in our synapses.

I think it is interesting that some emotional content or association is needed to move information into long-term memory. I said in the last part that feeling and emotion are the medium of our exchange with the field.

In the vast majority of cases, we get our own memories and not anyone else's, so our "local" minds must do a pretty good job of tagging the fragment we project with our unique identity. That insures that we only get back our own memories.

Chapter 11 — Thinking in the Field

What happens if we put out a fragment that is not connected to us. What if we are looking for information that does not belong to us? The answer appears to be that we get what we are looking for.

Beyond memory

Can we retrieve information that is not ours? If we consider the success that some psychics, remote viewers and medical intuitives have, we have to answer that we can get back information that isn't ours. I propose that the process of getting other information is the same as the process for getting our own memories: We project some fragment and we get back information that is associated with the fragment in some way.

Medical intuitives and remote viewers offer the most explicit examples of this process. A medical intuitive is given the name and city of some person. The intuitive focuses her intention on the name and information "appears" in her mind about that person and his or her inner state. The intuitive can convey that information in words. The name and city are enough of a fragment to allow the intuitive to make the association, or connection, with the information about the person and the inner state of that person. It takes no more effort than remembering your trip to Colorado when you were 12 years old.

Consider the range of information that is available to people:

- Medical intuitives read the health of distant people.

- Remote viewers see places distant from where they are physically located and sometimes places as they were in the past.

- Healers see physical and emotional problems in people and other living things.

- Psychics read the emotional state of other people.

- People doing past life regression and the children who just know things that happened in their previous lives and know things that happened to people in the past.

- A few people seem to do a pretty good job of seeing things in the future. Troubles and disasters are among the most commonly accessed future events.

What categories of information are available for distance access? In order to be available to people, the information must be distributed in the quantum field. We can look at the range of information available to people and conclude that light, i.e., the appearance of things, is certainly distributed in the field and can be viewed by anyone, anywhere. A person's state of inner health is also readily accessible and so must be distributed in the field. Strong feeling and emotion, particularly bad feelings and emotions, are accessible. These sorts of feelings are even available from the future. From this informal survey, I could conclude that light and feelings are distributed in the field, where "feelings" includes health, as in, "How do you feel?" and the feelings associated with mood and emotion.

I can think of lots of other kinds of information that does not appear to be readily available for remote access:

business and financial information, the ending of the Harry Potter books before the last book was published, test and final class grades, and most information that is not connected with strong feelings. I am making a distinction between the information contained in light about the appearance of a place and other kinds of information that people trade in. Every point on earth and beyond earth is available for viewing, whether it has feelings associated with it or not. Other human information appears to need some feeling or emotion associated with it to make it available in the field.

Making connections

It is possible to "see" any place where we direct our attention. The attention can be directed by almost any information that identifies the place. A person's name and city are enough for a medical intuitive. The coordinates of a place are enough for remote viewers. The web of associations needed to make this work is incredible.

Let's look at that for a minute. We have the giant hologram in the sky that contains all non-material information including, apparently, all the light, from the present, the past and maybe the future (more on the future later). People can retrieve a specific bit of that information using a very small tag and the tag does not have to be a direct fragment of the information being sought.

Remote viewing is the practice of seeing something in a location that is hidden or distant from you. It is practiced by many indigenous peoples. There was also a remote viewing program sponsored by the Department of Defense, DoD, at Stamford University run by Hal Puthoff and Russel Targ.[1] The DoD was concerned that the Russians were doing similar work. The remote viewers could use the latitude and longitude of a place. Those coordinates, which I am pretty sure had very little meaning to the people doing the viewing, were enough to allow the viewers to access the appearance of the place. The coordinates are a completely arbitrary grid which only make sense when you know where they start. While there are people in the world who know where the coordinates start and the approximate coordinates of many places on the earth, I'm sure it is not a large number of people, and I am pretty sure that the remote viewers were not among them. However, it appears that it is sufficient for the knowledge about the coordinates to simply exist somewhere in order for the remote viewers to use the knowledge to connect to the place at the selected coordinates of latitude and longitude. In order for the remote viewers to see the place, they have to access the body of knowledge about what the coordinates mean, that is, where 25 deg 13 min North, 133 deg 53 min West is on the globe, and then they can access the appearance of that particular place.

Medical intuitives use the same sort of non-local meaning. A name and a city are enough for the intuitive to connect to the target person. How does the field know that "John Smith" is a person and "Batavia, Ohio" is a place? In the tests of medical intuitives, someone usually knows who the target person is. The person who knows the target may not be present when the name is given to the intuitive, but the specific person who is the target is known. I wonder what would happen if the target person was not known, or was not specified exactly: as in "John Smith, New York, NY," for example.

Chapter 11   Thinking in the Field

People can follow multiple associations to get from the fragment of information they have to the information they want. Stated like that, the field sounds more like an active network than a set of passive associations.

I find that I regularly slip back into thinking about the field like a material storage room. I began by describing the quantum field as holographic memory. The holographic memory that the computer researchers are working with is "content addressable" in that a fragment of the stored pattern can be used to access the complete stored pattern. A simple physical matching of fragment with the stored patterns takes place. Now when I am talking about how the field works, I find I am still thinking about it as a physical pattern matching. The result is that I am amazed at how the field can "figure out" what 25 deg 13 min North, or "John Smith in Batavia, Ohio" means. It would be much easier to explain if the field dealt in meaning and not in simple physical patterns. What gets stored in the field is the meaning of the information, not just the data of the information.

Meaning

Today's computers store the data of the information. That's why computers are so dumb. Fast, but dumb. The field deals in meaning. That is why the field can easily return images of tropical islands to the remote viewer who said, "157 deg W, 20 deg N" and desert to the viewer who said, "15 deg E, 21 deg N".

I have said that it is feelings that drive our communication with the field. Feeling is similar to the notion of meaning, as in "What I really mean" is very close to "What I really feel." The people who figured out the latitude and longitude grid put the meaning of the coordinates into the field. After that, anyone else, for example remote viewers, can access the location just by using the meaning of the grid. They don't have to know the details of the information.

The idea of content addressable holographic memory is just a metaphor for how the field works. The field is actually a meaning addressable memory.

Creativity in the field

A large fraction of the instances of "knowing" things, that is of receiving information, is concerned with old information or information that is already there, like the memory and remote viewing examples in the previous paragraphs.

People and animals, however, can receive completely new information from the field. I am talking here about creative and problem solving information. Writers, poets and composers very commonly report that they are not creating anything. They are just writing down what is coming through them from some outside source.

I recently heard a talk by Elizabeth Gilbert (author of *Eat, Pray, Love*) titled "Nurturing Creativity" (www.ted.com). She described talking to Ruth Stone, a poet who was in her 90s when Gilbert talked to her. Stone described working in the field (the kind with grass) as a child on her family farm in Virginia. She could hear and feel a poem coming toward her across the landscape. When she heard a poem coming she had to "run like hell" to get to a paper and pencil before the poem got to her. Then she could write it down as the poem passed through her. If she didn't get to the paper

Chapter 11 — Thinking in the Field

fast enough, the poem just went right through her and continued on "looking for another poet."

Mozart did not compose his music with conscious effort. He just wrote down what was coming though him. For example, between June 26 and August 10, 1788, Mozart composed the 40th symphony and completed the 39th and 41st symphonies as well – a prodigious amount of music. These works are among his best.

Not many of us are prolific writers or composers, but I think that almost everyone has received solutions to problems "out of the blue."

The usual scenario goes like this:

You have some problem and you are struggling with what to do. You work and work and then give up, take a break or go to bed. Sometime later, in a dream, when you first wake up, in the shower or while you are taking a walk, poof!, the solution pops into your head. It is a solution to your specific problem. It is not a solution to your neighbor's problem, nor to the problem you had last month. It is a solution to your current, active problem. The common feature here is that the solution appears when your conscious mind is disengaged. This is exactly the time when we are most open to intuitive inputs, that is, to receiving information from the field.

In some cases, the solution is truly new information, not a reworking of existing knowledge.

This is an indication that the information is coming from "outside" the individual. It is common for new discoveries in science to be reported by several people at about the same time, even when they did not know there were others working on the same problem.

How does this work? The first part works like memory. A person broadcasts a need by thinking about and having feelings about a problem or a need, like "I want to write a symphony." In this situation, the information being sought does not exist. The fact that a solution often pops into the person's head tells me that the field is working on it. The fact that the same solution can appear to different people confirms that the field is doing the thinking and creating.

The field thinks or feels

Up to this point, I have described the field as a repository of meaning, which has a rather passive and inert sound to it. Here I am suggesting that the field is thinking on its own. But we already knew that.

Earlier I said that our thoughts, feelings and everything else originate in the field. Our material bodies are just the projectors of information that originates in the field. What's new here is that I am suggesting that the things that are thinking in the field are bigger than mere humans. I say the thinking entity is "bigger" than humans because it appears to be smarter than humans. It is able to create new things that have never been seen on earth. The fact that multiple people can tap into the solution at about the same time confirms for me that the source of the creativity is at least outside any one physical individual. I will take up these larger entities in the field in the next part.

If the field thinks on its own, what exactly does it think about?

Chapter 11　　　　　　　　　　　　　　Thinking in the Field

The problem with Symphony 4 in G minor for flute, oboe, clarinet, bassoon, horns and strings

People get help with their particular problems. Mozart got the G minor symphony, Elias Howe got a sewing machine needle, mystery writers get mystery plots, poets get poems. Does this mean that the field is busy composing a symphony in G minor and orchestrating it for flute, oboe, clarinet, bassoon, horns and strings? To my small mind, that seems somewhat improbable. But then where did the G minor symphony come from? I think that Pribram's perception model might provide an answer.

Recall that in Pribram's perception model, we take in holograms of various aspects of our sensory inputs. Those little holograms are sent out to a sensory memory. The remembered holograms are collected and sent back out the sense channel where they form the perceived image or information. What we perceive is actually fabricated from our own past experience. I propose that our perception of non-sensory information works the same way, mostly. The difference is that for non-sensory inputs there is no sensory memory directly associated with it so everyone learns to use a different kind of memory in which to wrap non-sensory inputs. Musicians use instrument voices and musical elements; poets use word structures and novelists use plot and character.

If people provide the details, what does the field provide? I believe that the field provides "structure and order" that relates to the problem being addressed which we interpret as feelings. It is not the specific form of the solution, like a symphony in G minor. This arrangement feels better to me. The field is "thinking" in abstract structure and order. People project the structure and order of their problem into the field and the field responds with a related structure and order. The human then wraps the received, abstract structure and order in the concrete elements of the domain he or she is working in. The result is a new creation.

If the field is so smart, why haven't all the problems been solved?

The field can create things that surpass the current capabilities of humans because the intelligence of the field is bigger than the intelligence associated with a single individual. The field also has access to everything, everywhere, so synthesizing solutions to specific problems is easier at the field level than at the human level. So why don't we know the solutions to all problems?

The answer lies in the mechanism that we use to perceive inputs from the field. We have to wrap the abstract structure and order that we get from the field in terms and concepts that we already know about. If we don't have the concepts needed to interpret the inputs, we simply can't see them. Recall the story about the indigenous people in South America not seeing Darwin's (or Cook's, depending on which version of the story you like) ship when it anchored in their lagoon. They had no experience with anything that big on the water and so it simply did not register in their awareness. They had an advantage over us as we try to "see" what the field is telling us. Darwin's ship was a material object and they could go out and touch it, walk on the deck and look in the hold. They had the opportunity to learn how to see a large ship. The non-sensory information from the field flows through our perception machinery. If we don't understand it, the information just looks like a strange dream and it's gone. Humanity has to wait until we are ready to perceive the information before what we receive makes sense.

Talking to the field

Up to this point, I have been describing people who are working on some project or problem and who receive a solution. The solution "pops into their heads" and they write it down or act on the information somehow or other. There is another way to interact with the intelligence of the field that is very common and that is to talk to it. Sometimes it is just words in your head, sometimes there is a felt presence conveying the words and sometimes there is a visual presence speaking the words. The conversations range from a simple question and answer to a long, involved discussion.

What do people talk to? Almost everything. It can be relatives (living or dead), spiritual figures (god, Christ, angels), aliens, fairies, animals, trees, rocks, inanimate objects. One of the features of this phenomenon is that when people perceive an entity and talk to it, they are absolutely convinced of its identity. It is not fruitful to even suggest that the thing they perceive to be there is not there, or that it could be some other entity.

The information seems to arrive from the field whether the person perceives a presence delivering the information or not, that is whether God is delivering the information or the music just starts playing in your head. That kind of thinking leads me to suggest that the presence people feel is also part of the local perception process. On the other hand, there is vast experience with dead loved ones appearing to people who perceive the vision as completely real in every detail. Besides, I like the idea of humans and other intelligences having independent existences in the field.

The information that people receive is often good, useful and sound advice. Although sometimes it is very bad advice, as when someone says, "God told me to kill my children." We'll talk about negative messages in a minute.

How does this work? The mechanism of this interaction is the same as for the other instances of non-sensory knowing. The feeling and meaning of the question is translated to the field via the microtubes in the neurons. The meaning hologram resonates with a matching pattern in the field and the field returns related holograms to the sender, just like the previous examples. The thing that is interesting about conversing with the field is that it demonstrates that the "information" is not passive data. It is a living intelligence. It can react and respond to inputs from the material person on the other side.

I have suggested that the "information" part of living things exists both before and after the material body's lifetime. That information part is not inert information. It is the living, dynamic part of our existence. It is the animating force for the material body while that body exists. Before and after the material body, the animating force exists as a living intelligence without a material body to animate. I have also said that if human "intelligence" can exist without a material body, then it should be entirely normal for intelligent entities to exist without ever having had a material body. So there are a great many intelligent entities in the field to whom those of us who are inhabiting material bodies can talk.

Chapter 11 — Thinking in the Field

I don't believe that the field talks in English or Japanese or any other language. The field works in feeling and meaning, structure and order. Talking to the field happens when we choose to send the intuitive information out through our verbal sense memory. We also provide the details to make the information specific to our situation.

Positive and negative messages

This wrapping process, where we wrap the meaning we receive from the field in words and images that have meaning for us, is sufficient to account for the negative messages that some people receive. I personally like the idea that the field always provides positive information. I am not alone in this reference. The idea that god is good is old and widespread, but there are lots of people who report hearing very negative messages. When a person has a very negative view of the world, everything is perceived as being bad. This means that the person's sense memory is full of bad interpretations. Whatever is perceived through physical senses is wrapped in negative interpretations. So, it is not too surprising that if inputs are perceived from intuitive senses, those inputs are also perceived as negative. We don't need to invoke a negative deity. We really do create our own realties.

One underlying mechanism

I have described what may seem to be several different ways of interacting with the field: retrieving existing information (as in memory), getting creative, new information (as in symphonies or inventions) and talking to the field or entities in the field (as in messages from god or recently departed Aunt Mim). I think there is only one mechanism for all of these interactions.

There are two parts to the mechanism. One is the process of wrapping the abstract order and structure we receive from the field with details familiar to us. Those details are drawn from memories associated with one of our areas of activity. The other is that the basic fabric of the field is intelligence. (You should probably replace "Intelligence" with "feeling", too)

If the idea just "pops into our head" then sensory memories are not involved. This accounts for things like simple memory recall or getting ideas for symphonies, poems and solutions to business problems. The information from the field is run though the domain knowledge of the person receiving the information: composition for Mozart and poetry for Ruth Stone. If the person hears or sees the information then he or she is running the information through visual or auditory memory. If the person has a conversation, then the information is processed through verbal and conversation memory.

All of these interactions depend on the field being an active intelligence. It interprets the meaning of the feeling holograms sent out by individuals and makes the appropriate connections.

The problem of bad ideas

People get all kinds of ideas-images-visions for all kinds of music, novels, businesses, solutions to problems, things to say, etc. If we look back and examine how those ideas turned out, we will see that a few of the inspirations turn out to be good and most of them range from OK to disastrous. Up to this point, I have been describing inspirations that turned out well, but those kinds of ideas are clearly not representative of the majority of inputs received from the field. So the question is, "If the field is so smart, how do we account for the dumb ideas?"

If we want to preserve the idea that the field is always good and smart, then we can propose that the bad ideas come from bad interpretation or bad wrapping during the perception process in the material intelligence. I think that bad interpretation is certainly possible and is likely quite common. Just look at the many ways different people interpret the information they receive in our material reality. What we perceive from any sense, including non-sensory inputs, is an active construction based on the raw inputs wrapped in ideas from our own sensory memory. What is in that sensory memory depends on how that particular person has interpreted events in the past. That mechanism is entirely adequate to account for all of the "bad voices" that people claim to hear.

I think there is another factor at work in producing the "bad" ideas, and that is that the field is a quantum intelligence. Like all the other quantum effects we know about, I propose that the intelligences in the field are also probabilistic and are not deterministic. This means that the information the field returns in response to the same request varies according to the probability distribution associated with that request: it might be "good" and it could be "bad" and everything in between. We will see in the next section on influence that the only thing that we can exert some influence over is the shape of that probability curve.

Life and Spirit in the Quantum Field

If true, this means the intelligences in the field are not as logical, clear-headed and brilliant as we might like to believe. This will be very important in the coming section on spirituality. If the field is god, then god is not completely "good." The field, then, is more like humans in that regard.

Intuitive insights range from good to bad, from brilliant insights to perfectly awful ideas. With a little clear thinking (not a common quality), it is usually possible to sort out the morally and ethically bad ideas from the good ones. It is much harder to sort out the brilliant insights from the crazy ideas. From my observations, it appears that the only way to tell the difference is in how they turn out, that is, by looking back at how the ideas were implemented.

That is the catch in using the gifts of insight we receive from the field. After we get the brilliant idea, it still has to be implemented in the material world, and the material world works just like Newton said it does: Things remain at rest until acted on by an external force. Worse that that, the choices and behaviors that we use to operate in the material world are processed through the same world view that I just described as being able to scramble even the best information we receive from the field.

There is a name for that part of us that wraps our perceptions in negative images: the shadow. It is an important idea in psychology. It shapes the feelings and expectations that we project into the field. For that reason it is a powerful force in determining what happens in our world. If you look at the practice section in the appendix you will see that addressing our shadows is very important.

So here we have the world we live in. There is an infinite source of intelligence "out there" that is accessible to everyone. It will answer our questions and help solve our problems and meet our challenges. The infinite intelligence is not, however, infinitely correct, good or right. It is a quantum intelligence so it is not always "right" in the way we would like. Add to that our own contributions to creation which can be everything from wonderfully creative to horribly negative. It is an interesting universe.

The advance I see in all this is that the source of the intelligence is real in scientific terms. That's important for those of us who are card-carrying members of the rational, scientific, Western culture. It means that the effort required to connect with the non-material intelligence of the field is a good investment because there are great benefits to be had. It tells me that the primary impediment to accessing and using the intelligence available in the field is me. As Pogo said, "We have met the enemy and he is us." The problem is our inner baggage, the subconscious stuff that colors our perceptions and actions, makes it difficult for most of us to use the information we receive in a positive way.

This chapter has been about knowing things. That sounds like a passive retrieval of information. We quickly saw that there is nothing passive about information in the field. That blurs the line between the topic of this chapter and the topic of the next: influencing. Knowing is really the same as influencing.

---

[1] Targ, R, Remote viewing replication evaluated by concept analysis, *Journal of Parapsychology*, **59**, pg 271-284, 1994.

Chapter 12
Human Influence

Back in Chapter 10 on influence, we saw that influence at a distance is real. We looked at the difference between the power that people desire and the limited scope of the real power. We also noted the mixed messages our culture sends about action at a distance: condemned by the cultural and scientific mainstream, but desired by almost everyone, the stuff of wishes and fantasy.

In this chapter, I want to examine what this ability means to us as we go through our days. The truth lies somewhere between the condemnation and the fantasy.

Health and healing

Our health is important to each of us, right? Everyone wants to be healthy and have everything work the way it is supposed to. It has been known for a long time that 60% of our heath is controlled by lifestyle. So it is rather curious that the majority of people in the United States maintain lifestyles that are clearly not contributing to health. Then we go to doctors who treat the symptoms with chemicals. I expect that many readers of this book will cluck, cluck at that, but the doctors would quickly be out of business if they prescribed lifestyle changes instead of pills. People do not like to contemplate changing their lifestyle. This is the background to our discussion about influencing health.

We now know, if we didn't before Chapter 10, that our physical, mental and emotional health is entirely malleable, that is, it can be changed with the force of intention and attention. We saw the reason for this in the previous sections. Our physical health and physiology are projections of the quantum-level processes that run our bodies. These processes can be influenced by the quantum-level processes of our thought and feelings. The result is the well-known and well-documented ability of people to heal themselves and of people to heal other people.

So, healing at a distance is possible and the mechanisms we have been developing tell us how it works, but it doesn't "work" all the time. Everyone doesn't get better. I would like to look next at what our mechanisms tell us about the limits of healing.

Self healing

Our health is a quantum process, so it is guided by the probability curves for health events. There are usually many other quantum processes involved that can exert an influence on the shape of those curves. Once the shapes of the probability curves are determined, then the outcomes of the events governed by those curves can fall anywhere under the curves.

People can move the curve enough to cure the negative symptoms in some people's lives, while others still suffer and die from their dis-ease even with the intervention of positive intention. I think it is safe to say that the most influential forces shaping the probability curves for a person are the feelings and expectations of that person about him or herself. Here's a little stronger way to say that: All healing is ultimately self-healing.

This fact about self-healing leads to a common bit of wisdom dispensed in energy healing classes: What you can do for others is the same as what you can do for yourself. Our innate ability to heal others uses the same mechanisms that we use to heal ourselves. As long as our ability to coherently focus our intent is intact, we can exert a healing influence on ourselves and on others.

The good news is that we are our own best influence for healing. No guarantees, nothing definite, of course, but the cost is low and there are no side effects or drug interactions. The bad news for most people is that we have to do it ourselves. Even when other people are working on us, we should at least avoid making a negative contribution to our own recovery.

This is the biggest challenge to using the power available to us: It depends on our inner state, including our subconscious inner state.

Pain and wholeness

There are paradoxes built into using our healing powers, at least for those of us coming from Newton's world of deterministic science. I would like to take up one of them now: the difference between healing and curing.

I have been talking about healing and healing that works. I expect that everyone knows what I mean. Healing that "works" makes you feel better. It removes the pain or other negative symptoms. It restores normal function. I think that everyone who enters a healing practice, from neurosurgeons to shamans, wants to help restore function and eliminate their patients' pain. There is another word that describes making things right, and that is "curing". Curing, as it is usually used, means removing negative symptoms and restoring normal function. In healing circles, curing is what allopathic doctors try to achieve. Healing is what healers do and it usually means "making whole." As we will see, making whole is not necessarily the same as curing, which means that it is possible to be healed and to still be in pain or even to die.

So we seem to have a paradox. Being healed and whole is an inner state that puts no requirements on the outer, physical state. Being a healer because you want to reduce your neighbor's pain requires that you really wish people to simply be whole. The way to do that is to accept the people as OK exactly as they are, in all their pain and imperfection.

The paradox can be resolved by observing that this inner state of acceptance of the pain and imperfection of the patient, which is the optimum state for healing, is also the best state for curing. That is, you get the highest percentage of cures using an inner state that accepts the imperfections of the patient. Keep in mind that "highest percentage" is a very relative term. It does not refer to anything like 80 or 90% cure rate. The cure rate is just higher than if you held a different kind of inner state. To use language of probability curves, seeking wholeness is the most effective way to move the probability curves toward health. That is getting us into spiritual issues, which will have to wait for another chapter.

Negative influence

If people can influence other people and themselves for the better, is it possible to be a bad influence on self and others? History tells us that the answer is "Yes." Curses, hexes, jinxing and all those voodoo rituals have a long history in human experience. Larry Dossey has three books out on the effectiveness of prayer. In *Healing Words*,[1] he describes the death prayer used in Polynesia to eliminate destructive or dangerous people. Keep in mind that when I talk about negative influence, it is subject to the same quantum probability limitations that constrain good influence.

How does this work? I went to some length in the last section to describe the very positive inner state, centered awareness and acceptance, that is optimum for conveying positive influence. So what is the inner state for conveying negative influence? First, positive and negative are relative terms, value judgments. It may well be that the death of a psychopathic individual would be the greatest good for the community. Outside of those situations, what of real, sincere negative intent?

There has been considerable work done on the effects and efficacy of healing intent. There has been very little, that I know of, work done on negative intent directed at other people. It is hard to say how effective such intent is. There has, however, been a great deal of work done on another kind of negative intent: the negative intent that individuals direct toward themselves. I said earlier that what we can do for others is what we can do for ourselves. If that is true, then all of the work of the psychologists and medical people regarding the effects of stress, shadow and negative self-talk becomes relevant.

It is well known that people can and do cause endless trouble for themselves due to the negative beliefs they hold of themselves. Many of these beliefs are tucked away in their shadows, so there is little or no conscious, intellectual involvement in them. Those negative beliefs are pure feeling. That makes them ideally suited to exert influence through the field. The target for those feelings is the person holding those beliefs about him or herself.

How effective is negative intent? On the one hand, it is capable of producing serious and often fatal diseases in people. We are talking about things like autoimmune diseases, heart failure and cancers. On the other hand, it seems to take many years of holding that negative intent before the physical disease appears. People have to work very hard at holding those negative images about themselves for a very long time in order to produce a serious disease. That tells us that the negative intent is a subtle but ultimately profound influence.

Using the equivalence principle (that is, what you can do for others is what you can do for yourself) it appears that causing negative outcomes in someone else's life should also take long and concerted effort on the part of the influencer. I expect that one ritual, or one or two sessions of directing negative intent at someone will not be very effective.

So is it possible to exert a negative influence on people? Yes, although we don't have to worry very much about other people's curses or voodoo rituals. We do need to be very careful about what we are doing to ourselves. We are all fully capable of exerting fatal negative influence on ourselves, and lots of people die every year from conditions significantly influenced by their own negative inner states.

Healing is a form of influence where one person influences the health of another. When we think of influence, it is usually in terms of one person affecting some specific target. Next I would like to consider forms of influence that involve whole webs of people and their actions.

Synchronicities

Many of the things that people want in this world involve the decisions, choices and actions of other people. I'm thinking of things like getting a job or promotion, being selected for a grant, scholarship, admission to a school. These sorts of situations might involve a few people, the people on the selection committee, for example. It is possible to add our influence to the other influences at play in those situations. This involves influencing a small number of people. There is another sort of influence that can involve many people. These are called synchronicities or synchronous experiences. Consider this example.

I talked to a woman who described traveling by car at a time in her life when she had very little money. The tires were bad and they gave out in a little town. She didn't have the money for new tires. Somehow she met a man who had four tires of the right size that he wanted to get rid of. The odds of those events lining up by pure chance are extremely small. But these sorts of experiences happen with some regularity. It is called a synchronous event. Your local bookstore offers many books on the topic.

Life and Spirit in the Quantum Field

For our purposes, the interesting thing about synchronicities is the vast web of connection and influence that must exist in order for them to happen. Consider the situation where you meet someone by chance and one or both of you have just what the other needs. There were many decisions and actions needed to bring you both to the same spot at the same time. There were many, many more decisions that went into each of you having just what the other needed and needing what the other had. If all those choices were isolated choices, the probability of that synchronicity taking place would be vanishingly small, but those kinds of events do take place in everyone's life.

By now we know that the choices are not isolated events. They take place in a sea of connection, which we now know to be the same as influence. Connections between individuals and their needs via the quantum field can be made based on the meaning of what they are doing and feeling. It is obviously not necessary that you think about "Joe Smith," who you will meet in a small town when you need new tires, before you meet him. "Needing tires" and "having tires" is enough to make the connection between those two people. The web is effortlessly expanded, as when Joe is in the diner where you are having breakfast because his 9:30 meeting with Jane Moore was canceled because she came down with a cold the night before. Then there are the string of choices that led you to be traveling with bad tires and Joe to have tires that he did not need.

The choices that we make are quantum-level events that are subject to quantum-level influences from sources anywhere in the world.

All the time

When I first became aware of healing and synchronicity, I thought of those events as unusual. I, or others, would have this transcendent experience and then we would go back to normal. The connected experience was fundamentally different from the normal way of being. As I began to delve into the mechanisms of how these things work, it became apparent that the connected experiences could not be different from my regular life. Every thought/feeling we have is put out into the quantum field where it automatically interacts with all similar and related thought/feelings. Our every thought/feeling, every choice we make is made in a sea of connection. Connection is normal. Isolation and separation are the illusions. Our inner lives, which drive our outer actions, are part of a web of associations that can include everything in the material world and a great deal of the non-material world of the quantum field.

Connection is not the unusual event; it is the normal way of existence. The thing that is hard for me to wrap my mind around is the scope and scale of the influence that we exert. I can understand doing healing work on one person, but the idea that my all-the-time feelings are out there pushing untold numbers of probability curves is amazing.

This, dear reader, brings us to the biggest lesson I have learned on this journey of mine. We have this vast, profound but subtle influence over what happens in our world. What kind of influence are we sending out? For most of us, our influence includes a large dose of stuff we like to think we are keeping hidden.

Our shadows and influence

The shadow is a term from psychotherapy. It is the place in our inner being where we put stuff that is not acceptable to our public persona. Some of it is normal drives that are not allowed in our family or culture, like lust or jealousy. A lot of it is beliefs (feelings, really) that we acquired in childhood about our value, worth and place in the world. Often these are very negative beliefs, distortions formed by young children faced with threats, real or imagined, to their physical or emotional safety. They keep us from being what we really are. The shadow feelings drive our behavior, which is why bringing the shadow into the light of awareness is so important in both psychotherapy and spiritual practices.

Our shadows are a bigger problem than just our mental and behavioral health. The problem is that our shadows, especially the negative feelings about ourselves, are pure feeling. Many of the feelings are preverbal so we never describe them with words. They never rise to our conscious awareness. Our negative feelings are perfect "messengers" to the field. They make up a big part of the influence that we exert. As we saw in the previous section, that influence extends far beyond our personal lives. It touches vast numbers of other people and events. If we carry around an unexamined shadow we are polluting vast reaches of the quantum field with negative forces that are not even true.

This situation is especially challenging because of our history. Our culture has dismissed and repressed feeling and emotion as weak and feminine for millennia. We have spent the last 400 years or so trying to believe with our conscious minds that we are isolated lumps of matter and energy with no connection to anything beyond ourselves. Descartes' deal is still widely observed: There is only minimal connection between mind and body.

Claiming our true power and exercising it for our own good is a big change to the way most of us lead our lives. It challenges our individual and collective egos, our self images, gender roles, our relationships to self, family and culture.

If all of this sounds like a major revelation, you should remember that I have simply climbed the mountain and found the sage already sitting there. The ways of being that are implied by all these quantum mechanisms have been taught for a very long time. As we can all see, the teachings have not caught on with very many people.

I hope that by climbing the mountain using the paths of science and power the path may be more appealing. I'm basically an idealistic optimist.

I will describe some of the practical work needed to climb this mountain of subtle power in Part 5.

Influencing matter

The quantum-level processes of our feelings and intentions do exert an influence on quantum-level processes in our material world. This is troubling to Newtonians and Cartesians (followers of Descartes), but the effects have been documented and measured extensively. Lynne McTaggart has a nice description in *The Field*.[2] Dean Radin gives examples of the data in *Entangled Minds*.[3] Both books have many references to the original work.

Life and Spirit in the Quantum Field

The main appeal of these phenomena for me is the challenge it poses to the mainstream scientific paradigm. Perhaps you recall that "mind over matter" is not allowed. That is not, however, the topic of this chapter. We are looking at how our powers of connection and influence affect our everyday lives.

What are quantum level events? The path of an individual BB down through the pegboard in Chapter 5, the outcome of a single roll of a pair of dice, lottery number drawings, and the random event generators described in the McTaggart and Radin books are all examples of quantum-level events. Before you start planning a life of ease from all the money you can make by influencing cards, dice and lottery numbers, I should remind you of a couple things. One is that the effects are subtle. You are not going to roll 7's 30% of the time. The other is that there are usually other players involved. They are exerting their influence consciously or unconsciously just like you are. The result is that it is hard to make money gambling. The results come out discouragingly close to random. Don't quit your day job.

Divination may be another area where we exert influence over material events. In this case the events are the "random" drawing of cards, sticks, coin tosses and the like. Divination is another of those persistent and widespread practices that indicates there is some underlying real phenomena. I think that part of why it works is that we can influence the random events of the draw to give us meaningful information. The other part of why it works is that we can see the future. We'll take up the future in the next chapter.

Quantum-level material events are very small. Influencing them seems to be much less interesting than influencing human health and behavior. There is, however, an interesting result that came out of the random event generator studies that I would like to consider: the effect of coherent groups.

Group Influence

Radin and McTaggart describe the work done at Princeton Engineering Anomalies Research, PEAR. The researchers at PEAR used random event generators, REG, to monitor the effects of human influence. The random events were generated by voltages from a white noise source (like the buzz between radio stations) or from decay of radioactive material. These events were randomly distributed over time. A computer monitored the events, tracked their randomness and displayed the events so they made sense to human observers. One display was a dial with a pointer that swung randomly to the right and left.

When one or more people directed their intention to "do more lefts" toward the event generator, the events became slightly, but significantly, less random. The computer monitoring the events tracked the deviation from random by keeping running averages over time. The deviations were not absolute changes, that is, all of the events did not suddenly start going one way.

The PEAR experiments started with individuals directing their attention and intention at one machine. They found that men did better then women at making the deviation go the way they wanted, but that women produced bigger deviations, just not always in the direction that they wanted. Couples, people in a relationship, produced bigger deviations than unrelated people.

Life and Spirit in the Quantum Field

The next stage of the experiments showed a different kind of influence. They took the random event device to a business meeting. They found that the machine deviated from randomness when the group came to a conclusion, that is, when they all shared the same view or idea about what to do. So, it looked like coherence of thought/feeling in several people could exert an influence on quantum level events.

Recall that back in the section on knowing things in the field, I suggested that the information people receive from the field was in the form of general structures and relationships. The person receiving the information is responsible for wrapping the general information in the specific forms of the problem they are addressing. This group coherence result in the PEAR experiments is the same phenomena, but in the other direction. Coherence and order in the material world (the people all focused on the same idea) can project influence into the quantum world, and influence random events.

Looking for other examples of group coherence to test, they took the machines to concerts. There, they found that at the climax of the piece of music, when the audience is most focused and connected to the music, the machine showed a significant deviation from pure random behavior.

How big is this effect? The largest scale they could test is the world, so they set up a network of random event machines around the world and monitored deviations from pure random behavior. They started with five machines and had 40 running continuously at one point. They looked for deviations that occurred on all the machines at the same time and then looked to see if there was something going on that might produce large-scale coherence of human thought/feeling. They found several examples of that coherence and influence. One was the Academy Awards television show. That show has a huge, worldwide audience. When the best picture, best actor and best actress awards were being announced, all of the random event machines showed a deviation from random. Another instance was the O.J. Simpson trial.

The news and television coverage of the 9/11 attacks on the world Trade Center produced the largest deviations recorded by the REG network. The interesting thing about that event is that the deviations peaked two hours before the first plane hit the building.[4] We'll take up premonitions in the next chapter.

In the examples of healing and medical intuitives, the practitioner thought about the person she was reading or healing. I described the connection as being the same as looking up something in holographic memory: You have to have some small key to make the connection. In the example I just gave, no one who was watching the Academy Awards or the coverage of the 9/11 attacks was thinking about the REG machines. Where is the connection if no connection was sought? I think that the answer lies partly in the simplicity of the REG machine. The REG machines are purely mechanical devices with a very small part –the part that generates the random events -- operating at the quantum level. Because it is a very simple quantum-level event, it can connect to, or be influenced by, any source of order in the field.

Or perhaps everything is influenced by the appearance of order in the field, from any source. Even human behavior could be influenced by every source of order. We would not be aware of the influence because it is so normal, so constant. It takes intent, directed explicitly at us, for it to register in our conscious minds.

I have said that in the field, thought/feeling is action. I also said that what people receive from the field is structure, order and relationships. The REG machine experiments show us that order, in the form of focused attention of many people, is a quantum-level force that can influence material events at a distance.

The implication of these group effects is that there is power in having a coherent group. The group coherence produces forces of order that cover the entire world. Order in the field is an influence for everything from health to symphonies. Cultivating order looks like a very good idea. I have to qualify that idea by adding that the order needs to be sincere and voluntary to produce the kind of coherence needed. Coercion does not produce healthy order.

Large scale effects

Up to this point, I have emphasized that quantum-level forces can influence quantum-level events. Quantum-level events are much smaller than the material events we see every day in our macro reality. But there are persistent accounts of large-scale effects. By large-scale, I mean peoples' intent apparently having an impact on macro-scale material objects and events. I'll give one particularly well documented example: the Jansenist miracles of eighteenth century Paris.[5]

The Jansenists were a Catholic sect that was unpopular with both the Pope and the French king, Louis XV. They were accused of being Protestants pretending to be Catholics. Their popularity persisted in spite of the

Chapter 12   Human Influence

official harassment because the Jansenists were very good at miraculous healings.

In May, 1727, Francois de Paris, a popular deacon in the group, died and was buried in the cemetery at Saint-Medard. His followers started gathering at his grave and soon many healings were reported. The Jansenist miracles were not these healings, but the convulsive state that overtook many of the mourners. In this state, people contorted their bodies into all sorts of unnatural configurations. They were also immune to all physical harm. They could be beaten, stabbed, cut and burned, all without any physical harm.

What makes the Jansenists interesting is the number of people involved and the length of time the miracles persisted. They went on for years and included a great many people and were witnessed by thousands. In 1733, six years after they started, the government was managing over 3,000 volunteers to assist the convulsionaires, as they were called. Voltaire visited and commented on the phenomenon. A member of parliament studied the convulsionaires and wrote four volumes describing them. He published the work in 1737.

Once again, I am taking the historical record to be a valid or a "close enough" indication of what actually happened. We have a phenomenon here that appears to contradict the normal laws of physics. Of course, everything we have talked about up to this point has also contradicted the normal laws of physics, but before now, we have been able to describe the events as quantum-level events. Here it appears that the quantum-level inner state of the Jansenists was able to influence material-level events, like sword strikes and rock blows.

How do we account for this sort of phenomena? Looking across many reports of this type of event, it appears that most of the instances of large-scale material effects involve great mental, emotional or physical stress or deep spiritual practice. The Jansenists may have had both. I do not believe that the people involved in these events have any sort of super, or even unusual, powers beyond those that are shared by everyone. If feeling and emotion are the medium of influence in the quantum field, then having a very strong emotion or feeling seems like a reasonable prerequisite for having a very strong impact.

We should keep in mind when talking about the material world that in the current quantum theory, there is no such thing as "solid matter." Everything, including the particles that make up rocks and swords, consists of weightless particles moving at the speed of light. They get the properties we associate with mass by interacting with the Higgs field, another mass-less entity in the quantum field.

The problem of influencing material reality with our non-material feelings and emotions, via the quantum field, is not one of stopping a massive sword or stone with a thoroughly non-material feeling. Rather it is one of getting a lot of non-material particles to behave a certain way using equally non-material forces of feeling and intention. I say this to indicate that the effects are possible. I don't want to indicate that it is easy, because it is obviously not easy. The phenomena are rare and usually involve extreme conditions, but it is possible. What does it take to make it happen?

The behavior of a collection of atoms, as in a sword or a stone, at the atomic level is very random. They are all vibrating in their own way. Atoms in a dense solid, like steel or stone, interact frequently with other atoms, so the group behaves in a very predictable Newtonian way. But the quantum level probability distribution of behaviors of individual atoms is still at work. Making a sword stop just as it touches the body of a Jansenist convulsionaire requires moving the probability curve into a very narrow peak for a great many atoms. This is seriously heavy lifting in the quantum world, which is why it is rare and requires extreme conditions.

What is the application of this sort of thing to our everyday lives? Given the extreme inner state that is required I don't think it is something that we need to worry about today or tomorrow. It does indicate that our relationship to material reality is very different from what we learned by banging our shins on chairs. A deeper understanding of this kind of "influence" could lead to a very different future.

The influence word

Our language grew up in the large-scale, material world so it is not too surprising that we don't have good words for the quantum field experience. Consider the word, "influence." The word sounds like something that one person or thing does to another. It implies a separation between the influencer and the influencee and this is not an accurate description of what we are dealing with in the quantum field. As is usually the case in this type of discussion, people a long time ago gave the matter a lot of thought. The Vedic creation stories fit very nicely here. In those stories, the universe was created when the primal sound, that is vibrations, was uttered into

Life and Spirit in the Quantum Field

the void. That primal sound is Om or Aum. The influence that creates the universe is sound waves. There is a modern theory of cosmology that suggests the same thing. Pressure waves in the quantum field were responsible for the formation of matter and galaxies in the early universe. Sound waves in the Vedic stories are a form of pressure waves.

I think the nature of what we have been calling influence is best conveyed with the imagery of sound and harmony. Sheldrake has used the term "resonance," a vibration phenomena, to describe the influence. Harmony is the name we give to resonance when we are talking about music. The waves that are interfering, resonating and harmonizing are the same waves that form the probability curves for the outcomes of quantum level events, so the harmonies are the influence. The quantum field is like continuous music, continuous singing. The simple act of forming a harmony with a sound is the "force" that changes the outcomes of events in the material world. It is not a "force" in the sense we normally use the word because it is not something that is separate from the things being acted on. In quantum reality, the "force" that shapes the outcomes in the material world is the same thing as the material outcome – but we have to add the qualifier that by material outcome we mean the outcome of all the events influenced by the same probability curve.

At this stage in our journey, we are the source of those waves and harmonies. Our inner state is continuously harmonizing (or making dissonances, as the case may be) with everything else in the field. It is really important, but not required, that we work on developing a harmonious inner state.

I have suggested that we may not be the only sources of sounds in the quantum field. We will take up the issue of the cosmic choir in Part 4.

But first, there is another capability that humans have that we can explain with the quantum life model. It is the ability to know the future. I'll consider that next.

---

[1] Dossey, L, *Healing Words,* HarperSanFrancisco, 1993, pg 154.

[2] McTaggart, L. *The Field*, Harper Collins, new York, 2002. Pg 111-122.

[3] Radin, D. *Entangled Minds*, Paraview Pocket Books, New York, 2006. Pg 146-160.

[4] Ibid, pg 203.

[5] Talbot, M, *The Holographic Universe*, HarperPerennial, 1991, pg 128-132.

Chapter 13
The Future

KNOWING AND INFLUENCING the future has always been a very desirable thing, probably for as long as there have been humans. There are rituals and ceremonies to ensure a good harvest or hunt. Our recorded history is full of oracles, dream interpreters and prophets. In a pattern that should be familiar by now, I am going to say that we can know the future because the future exists in the field. I am also going to say that the knowing is not as clear or simple as your local fortune teller claims, and that it is much more possible than the Newtonians and Cartesians among us would allow.

Everyone wants to know the future

The practice of consciously seeking information about the future is called divination. Divination practices use everything from tea leaves to chicken entrails to tell the future. The *I Ching*, a Chinese divination manual, was written down in its present form 3,000 years ago and remains in print and in use today. Nostradamus' prophesies remain popular and the Mayan calendar has attracted great interest. Premonitions and dreams about the future are common. You probably know people who have had premonitions that came true, perhaps you have had some yourself. There are a huge number of reported premonitions of the 9/11 attacks.[1] Weather reports and the leading economic indicators are very popular, if not terribly accurate predictors of future events. Psychics and fortune tellers persist, but you don't see many scales advertising "Your weight and fate, 5¢" any more.

So knowing the future is desirable. It is also one of those widespread, persistent phenomena that point to some underlying, real phenomena. The reality of the phenomena is an issue because it is not allowed in the Newtonian world view. Predicting the position of a satellite or a planet in the future is OK because it can be calculated from physical laws. Predicting whom you will marry or when you will die is not OK because there are no physical laws covering those events. There is, of course, science being done on the topic. Larry Dossey has a new book out on premonitions, *The Power of Premonitions*, that gives a nice overview of scientific evidence for knowing the future.

The future and free will

There is a bit of a problem with knowing the future if you think about it, although it doesn't seem to bother many people who are trying to know about the future. The problem is with predestination. The normal model is that in order to "tell" or "know" the future, the future must be "there" to see. So if people can see the future, it must be all laid out. We are just like trains on a track. We can only go where the track goes. We have no choice in what we do or what we experience. This view of the world is not very popular. Most people want to believe that they have "free will" and can make choices that make a difference in their own lives and in the lives of others. There is another problem with knowing the future that is a popular issue in science fiction. If you know the future, then it becomes possible to take actions in the present that will make sure the future happens differently than what you knew before you took the actions. If that is true, then what does that tell us about the fixed, predestined future that is lying out there for people to see?

The future in the field

Dividing experience into past, present and future is an artifact of living in a macroscopic, local and temporal world. In fact, having a past, a present and a future is the definition of a temporal world. I have described the quantum field as a non-local, non-temporal place where information, thought/feeling and probability waves live. I have also described the field as a hologram of our material world. This means that information about what we think of as "here" and "there" and "now" and "then," is spread out across time and space by the big Fourier transform in the sky. Massive, material things, according to the quantum equations, remain local and temporal. So if the future is knowable, the likely place for it to be is in the quantum field.

If the future is "in" the quantum field, what is it? It can't be my physical body doing its thing next week or next year. My body is local and temporal, isolated in the present, moving into the future with clockwork precision. The field, on the other hand, is information, thought/feeling and probability, so that is what the future must be.

Perceiving the future is then the same as perceiving any other non-sensory information and it is subject to the same vagaries and pitfalls of interpretation. I believe we get abstract information from the field. We wrap that information in forms drawn from our personal memories so we can understand what we have received. The process is the same as Mozart's process for writing all those symphonies.

What does it mean to say that, "the future is in the field?"

Describing a quantum future

Let's try this analogy. The material present is like a fish swimming through moving waters. The water is like the field. It is spread out behind, around and ahead of the fish. Only part of this analogy works. The part that works is that the fish's present is affected by forces and movement in the water that the fish encounters. Those forces do not control the exact movement of the fish, but they constrain and influence the choices the fish makes.

The part of the analogy that doesn't work is that the water in the stream is linear. The water downstream of the fish is separated from the water upstream of the fish. The field is all in the same "place" all "at once." Maybe the analogy would work better if, instead of a stream, we imagined a fish swimming in a hot tub with the water jets on. Now, all the water is in one place and the fish still thinks he is encountering a linear, sequential series of influences.

How does telling the future work? Here is my proposal. The future is made out of the same non-material stuff as the present, and everything else in the field. It is also "concurrent" with the present. The future can be perceived just like anything else. The key to the "fixed future" problem is that future waves are no more fixed than any other waves in the field. They are continuously resonating and interfering with other waves. The connection and resonances, and even the "just looking" events, all influence the future waves just like they influence the present waves.

Or, to use the music and harmony analogy, the sound of the singing resonates throughout all of the field, which includes the past, present and future of our massive, material world. Imagine an underwater loudspeaker in the hot tub with our fish. The sound travels through and fills the whole tub, which includes the past, present and future water that the fish swims through.

As usual, only part of that analogy works. The part that doesn't work is that in the field, the sound is not being generated by an outside source. It is being generated by the waves in the water. The waves are singing their own vibrations, which influence the singing of all the other waves, all of which exert subtle influences on the fish that swims through them. So the future in the field is just like the present. It is a sea of living waves. The waves are thought/feeling that resonate and interfere with other waves and when the time comes, the waves define the probability curves for the quantum level events in the material present. Our seeing or telling the future is just one of the many resonances that go on in the quantum sea.

Nature of the future

The structures and relationships that influence the future are the same as the structures and relationships that influence the present. They are all available, they are all influenced by every connection or resonance that touches them.

We can see the future more or less clearly depending on the clarity of our perception mechanisms. The question of whether we can change the future because we know it is not a meaningful question. Our existence as a set of waves in the field means that we are continuously changing the future, just as we are continuously influencing all things in the present.

We don't see a fixed future. We see the waves and currents in the water ahead of our material bodies. Like the fish, we are different from the water, but unlike the material bodies of ourselves and the fish, our non-material selves are the same water waves that our material body swims through. Because we are swimming in a tank of holographic waves, the waves we swim through are the past, present and future.

How Does It Work?

We remember the predictions that come true. We tend to forget predictions that don't come true, both our own predictions and those of others. The waves we perceive as the future are still probability waves. Material events (the ones you predicted) still unfold under probability waves. There is nothing deterministic about it. Even if you saw the waves "correctly," it is still possible for the events to come out in the tail of the curve and so be different from the prediction.

To successfully tell the future, we have to resonate with the information about the future situation. We have to wrap it in appropriate images so we can perceive it as a meaningful event in our lives. In the material time between our prediction and when the event unfolds in our material world, the probability curves for the event have to remain unchanged. With everything subject to change due to other resonances, the curves could become something different before they unfold in the material world. Finally, the events have to come out close to the "position" in the probability curve where they were originally "seen."

This is why there are relatively few correct predictions of large-scale events. The probability curves are necessarily "large," so there is room for lots of variation in the outcomes. There are also lots of people, or other influencers, directing their intent, but our free will, just like the future, is intimately connected with anything and everything that resonates with us, and that resonant connection influences what we choose. The exception to that statement about "relatively few correct predictions" is events with a large emotional impact on people.

Feeling and Emotion

I suggested in Chapter 10 that feeling and emotion are the basis for our interaction with the field. Spontaneous premonitions of the future provide more support for that idea. The most common subject of premonitions is trouble: disease, catastrophe and death.[2] There are some premonitions that are not about doom and gloom, but the great majority of spontaneous premonitions are. The medium of the field is what we call feeling and emotion. We need feeling to project our influence into the field and we get feelings and emotions back out.

I suspect that the distinction between premonitions that arise spontaneously, as in dreams, and intentionally seeking information, as in divination, is important. The information needs a stronger "push" in the form of stronger emotional content to penetrate to our conscious awareness when we are not expecting it.

The Future

We live in a hologram of material space and time called the quantum field. Time is distributed in the field along with space, so the future is accessible to us here in the material present. The field is a living and changing thing. Everything is subject to influence from many sources. Knowing the future does not imply a fixed future. It may be just the opposite: Knowing the future is a form of influence.

What is the value in our daily lives? The information we get spontaneously from the future is about coming problems. Cultivating our ability to perceive that information has considerable survival value. There were many people who chose not to go to work in the twin towers on 9/11 and who canceled their booking on the Titanic because of forebodings of disaster, or at least a hunch.

How far have we come?

We have considered mostly human possibilities and capabilities in this part. We are able to tap into information from anywhere. Thought/feeling takes place in the field, so the field is thinking. Our ability to resonate with order and structures in the field that relate to our own situation means we can tap into creativity in the field. Of course we have to be able to comprehend the information we get: otherwise, it is just "a crazy dream."

Chapter 13 — The Future

Since the waves in the field doing the thinking are the same waves guiding the outcomes of events, thought is action in the field. We can exert material influence with our thoughts and feelings. Most of the time, the influence is subtle, but on rare occasions people can alter the course of large-scale, material events.

Both time and space are distributed in the field. This is why people can and do access information about the future. The future is the field, so it is accessible and malleable like everything else in the field.

The unity of thought/action, past, present and future available to us through the field, is a quantum leap of understanding. It is also the basis of the amazing capabilities available to humans if they are willing to let go of Newton's model of reality.

Knowing and influencing across time and space are usually reserved for gods and other spiritual entities. We will see in the next part that, indeed, gods do have the same abilities as humans, although the scale is different. Or perhaps we should say that humans have the same abilities as gods. It turns out that "Ye are gods" is not an idle bit of poetic license. That is the topic of the next part.

---

[1] Radin, D. *Entangled Minds*, Paraview Pocket Books, New York, 2006, pg 29-34.

[2] Dossey, L. *The Power of Premonitions*, Dutton, New York, 2009, pg 114-119.

Part 4: All is One, Ye Are Gods

We have seen that we humans have some pretty amazing abilities. Some of those abilities, like knowing things and influencing things at a distance, are the kind of abilities we usually associate with gods, or maybe subgods. Even if that is true, there are very few people who would describe those abilities in spiritual or religious terms. As a culture, we do a pretty good job of keeping our spiritual lives separate from our secular lives.

There are some old ideas that hint the separation might not be vaid. Two that I like are "All is One" and "ye are gods." All is One is a bit of Eastern wisdom that is a popular bumper sticker in some circles. Ye are gods is from Psalms, 82. Christ mentions it in John 28. Both of these phrases tell us that humans are more than "mere mortals." We will see in this part that gods may be less than the stereotypical god.

My journey started out trying to explain how energy healing worked using the emerging quantum life sciences. The new models led to explanations of many things besides energy healing. When it became apparent that the quantum field combined living intelligence and influence, I saw a physical mechanism that could account for the entire spiritual part of human experience. It began to look as if the separation of the spiritual and secular parts of our lives into separate domains was just an artifact of Descartes' fear of the Inquisition. Here was a way for what we usually call spirit and only visit in church, to become a normal part of everyday life.

For those of you concerned about such things, the explanations and mechanisms preserve the unknowable god.

So join me on this next part of the journey. It gets really interesting.

Chapter 14
Mechanisms of Spirit

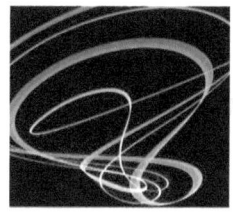
I HAVE SPENT MUCH of my life seeking spirit in science. I have also spent much time avoiding and criticizing organized religions. The avoidance part must have been working extra hard because I was surprised when my work explaining how energy healing works led me into how the world of spirit works. It felt like I had bumped into something in the dark. I felt a little silly when I realized that I had arrived at precisely the place I said I wanted to be.

My path to accepting the reality of spirit has been a little strange. Nonetheless, I feel that I am in a good place because I find myself describing some very old spiritual models and practices. After all, as Kepler and Galileo showed us, the best verification for a theory is agreement with observation and spiritual practice is nothing if not observation.

The spiritual domain

I would like to begin with the widespread and persistent aspects of the spiritual domain and describe how they work in the quantum life model we have developed here. Let's start with life after death.

Soul, ghosts, life after death and reincarnation

Life after death is a very common idea in spiritual traditions. There is something associated with a material being that does not die when the material body dies. This something is variously called spirit or soul. It is considered to be the animating force that gives life to the lifeless lump of chemicals that comprise a material body. The normal view is that when the material body dies, the soul or spirit leaves the body and has some sort of non-material existence, independent of the material body. The place where all these spirits live is called heaven or hell. We'll talk about that in the next section.

In some versions, the stay in heaven is permanent and that kind of non-material existence has many, varied descriptions: a halo, wings and a harp on streets of gold, 72 virgins, union with god. In other versions, the stay in the after life is short-lived and the spirit is given a new material body to live and learn again.

Today, reincarnation is usually associated with Eastern religions, but it was part of the Christian teaching until 553 when it was declared a heresy at the council of Constantinople. In traditions that actively teach about reincarnation, a process is described where the spirit gets a life review and perhaps some instruction and is sent back for another material life.

If all these souls live on past the death of the physical body, then they should still be "there" to talk to. Indeed, ghosts have been part of human life and literature for as long as we have any records. Visitation by recently deceased family members is a very common experience. Most near-death experiences include encounters with those who have already died.

How does this work? Perhaps I have given away the mechanism in the previous pages, but we can review it here. We think/feel in holograms of our sensory inputs. The quantum field is a hologram of our material world, and the waves have the same form as our thought. Actually it is the other way around. Our thought/feeling has the same form as the field. Our thought/feeling originates in the field. Our material bodies and brains are projectors for that information. Our particular patterns of thought/feeling define each of us as a unique being. So we can say that the "force" that animates our bodies is our essence in the field. That's one role that the soul or spirit plays.

We said that things in the field that have no mass are distributed across time and space, that is, they are non-local and non-temporal. Thought/feeling has no mass, so it is distributed. Being non-local means that our thought/feeling is "everywhere." Being non-temporal means that it is "everywhen." From the point of view of our very local, temporal material bodies, this means that our source of thought/feeling exists for all time. It is eternal, which means without time. This is a very nice match with the thing that people call the soul. It animates the body during life. It exists past the death of the physical body. It is the essence of the person, both during the life of the physical body and when the physical body has died. So I propose that the phenomenon that underlies the widespread idea of soul is our living, changing thought/feeling that is distributed in the quantum field.

There is an interesting implication of this model. If this thought/feeling essence of us is "without time" and the body is very much "in time," then the essence exists both before and after the limited time of the material body. Or to use conventional spiritual language, the soul exists before and after the existence of the material body. This is a little different than the Christian model where the soul is assumed to start with the creation of the physical body. It does nicely account for reincarnation. If we can develop one temporal material body on the non-material soul, why not more than one?

Heaven and hell

The normal view is that there is life and there is afterlife and they are different. There is the notion that when you die, you "go" somewhere. Indeed, the image of moving and traveling is common in the reports of near-death experiences. There is also the widespread idea that the quality of the experience in the after life is related to the experience in the material life.

If our soul is the non-material, non-local, non-temporal part of our thought/feeling, then it seems likely that the afterlife is closely related to the material life. People certainly have a huge variety of experiences in their lives on earth: from material poverty and fear to material luxury, from inner torment to inner serenity, from psychopathic behavior to saintly behavior. Each person's experience results from a combination of life events and the meanings and interpretations they associate with those events. We do create our own realities based on our beliefs about the events that befall us. We saw in previous chapters that our beliefs also play a role in determining what those events are.

It is our deeply held beliefs that appear to be what is distributed in the field. That being the case, we can then say that it is our deeply held beliefs that determine what kind of afterlife we experience. Indeed, our beliefs *are* our afterlife.

We don't have to go to heaven, hell and the afterlife. We are already there. We are always in heaven. What changes between having a material body and not having a material body is the nature of our sensory inputs. When we have a material body, much of the mind is occupied with processing inputs from the physical senses. When the material body dies, so do the physical senses. We are left with what we call intuitive inputs. I also assume that the perception process changes. We may not have to process the inputs through our local, sensory memory. That is, it may be possible to perceive the structure and order of the field directly.

The sensation of traveling, or going somewhere when we die as reported in many near death experiences, can be explained by the change in the dominant sensory input. The sensing of the material world stops and the direct perception of the content of the field begins.

There is an important implication for religious practice here: we are always in heaven. We don't have to go anywhere to enter heaven. Another implication is that suicide may not be much of a relief from some kinds of trouble because our essence is the same after we die as while we are living.

There are things besides souls in the spiritual world. Let's look at some of those entities next.

Angels, demons and other spiritual entities

There is a very interesting idea that follows from the mechanisms in the previous section that thought/feeling originates in the field and that our core thought/feelings exist before and after the material body. This means the field can think/feel and influence without a material body. This is only interesting if we are coming from the place that says our material body is real and all the inner stuff is not. Material minds are not the source of thought, so the material mind is not needed to think. It means that it is probably common for intelligences to exist that never have a material body.

If we could look at the field, we would see lots of human intelligences that spend most of their "time" without material bodies. Once, or once in a while, they acquire a material body and spend some linear time growing and living a life on Earth. After a short while they are back to being a pure intelligence in the field.

Long human experience tells us that there are other entities "out there." My suggestion is that those other entities are intelligences that never have material bodies. They operate perpetually in the form and structure of the quantum field. In that regard, they are very like the intelligences that do have material bodies once in a while, like us. The fabric of the intelligence for these entities is the same fabric that makes our intelligence. But how might they differ? I think the difference is scale.

The issue of scale

What is the "scale" of an intelligence? Even living things below humans have feelings, the same feelings that humans have, and those feelings exist and persist in the field. So we have the notion of "intelligences" that exist in the field that are, somehow, below the scale of human intelligence. Then there are humans and human thought/feeling. I like to think that my thinking is bigger than that of a worm, for example.

We can say that there is a spectrum of intelligence in the field from whatever the smallest is up to human intelligence. All of these intelligences are distributed across time and space in the field, which means that they exist before and after their material bodies. It seems entirely reasonable to say that intelligences can exist that are further "up" in the scale spectrum than humans. They are bigger than human intelligence. The fact that they never occupy a material body is not an issue because even intelligences that are associated with a body spend some, or most, of their time not connected with a body.

These intelligences that are never connected with a material body are the basis for the persistent images of spiritual beings of various sizes, like angels, archangels, demons and sub-gods.

These intelligences do not have material bodies, so any body that they have when perceived by people is supplied by the observer's memory and perception process. Another hint that the bodies are supplied by the observers is that the bodies vary from culture to culture: angels in the West, dragons in the East.

What, exactly, is the difference between a "larger" intelligence and a "smaller" intelligence? In material brains, the difference is indicated by the depth of structure of the brain. Insect brains are simpler than lizard brains. Mammalian brains add a limbic system to the lizard brain. Human brains add the cerebral cortex to the mammalian brain. But there are no material brains in the quantum field. So what differentiates the levels in the field? I would like to propose that it is the scope of awareness. A worm can sense the dirt around it. I am pretty sure that worms have non-local connections just like everything else, but I don't expect they have many conscious thoughts about it. Humans can be aware of trends and events in their own community, in communities of lesser animals and in the environment, including the stars in the sky. Humans can perceive non-local information about any of those domains.

If we use "scale of awareness" to distinguish levels of intelligence in the field, then the "larger" intelligences can see more than we can. Of course, these higher level intelligences don't have bodies, so they are not seeing the material world that we live in. They are not limited by having to process a lot of local temporal sensory input. Their scale of awareness is defined only in the non-material quantum field, the world of influence, connection and probability. If that is true, then humans and other intelligences that sometimes have local temporal material bodies must have that kind of awareness when they are between bodies.

What are some examples of scales of awareness? I mentioned the difference between what I presume a worm is aware of and what I am aware of. Consider variations in awareness among people. You probably know someone who is completely self-absorbed. He is exquisitely aware of his own problems and issues and seems to be completely oblivious to the people around him. Consider people whose awareness is limited to some group: tribe, town, religious sect, social class, culture, race, or any other division. People in their group are automatically OK and everyone else is at least suspect if not down right evil. These divisions can be described in terms of scales of awareness.

A little later, I will talk about scales of awareness in non-material beings. Looking ahead to those examples, the awareness that manages the formation of a snowflake is a "small" scale of awareness. The awareness that manages the evolution of all giraffes is a larger scale awareness. The awareness that managed the Big Bang is as large as it gets.

I should note that in the quantum world, seeing or awareness is the same as influence at the quantum level. So the scale of awareness is the same as the scale of influence.

I find that to be an amazing idea. Since I am writing in my material body state, it is hard to imagine what that would be like. However it is, entities that never have bodies live like that "all the time." Of course, time has very different meaning in the non-temporal quantum field.

Humans in their material phase (that's us) can apparently perceive something of these non-material entities. I take the long history of human experience communicating with various non-material entities to be sufficient supporting evidence. That means we can "see" a few levels up the hierarchy of intelligences. Because the medium of existence for all entities is the same (the living, feeling, influencing waves of the quantum field) we can communicate with these beings. Actually, the phrase "can communicate" is not right. That makes it sound like we communicate when we choose to and we don't when we choose not to. That's not the way it is. Given the nature of the field, particularly the holographic memory aspect, connection and communication occur whenever any entity thinks/feels anything that has any common meaning with anything else.

When we do perceive information from these higher intelligences, most people do what they normally do with perceptions, which is to wrap them in images drawn from their personal memories, which are usually determined by the culture in which they grew up. So people in the West see angels and people in the East see dragons. Most non-material things are considered to be in the spiritual domain, or more accurately, most non-material things that we believe to be of a higher level than ourselves are considered to be in the spiritual domain.

God

God is a touchy subject because it is so laden with rules and prohibitions, both in religion and in science. My belief is that if science is supposed to tell us about how the universe works, and if god is a real part of the universe, then we should be able to use science of some sort to talk about god. It is possible with the models we have been talking about, so here goes.

I described a hierarchy of intelligences in the previous section. That hierarchy was based on the scope of awareness of each entity. What then is god? We like to think of our god(s) as the ultimate power, the ultimate creator. If we go with that definition, then I can say that god is the quantum field for the whole universe. The field is thinking/feeling, connecting and influencing, so the ultimate force is the whole field.

Chapter 14 — Mechanisms of Spirit

There is another bit of old and widespread wisdom that tells us that god is unknowable, ineffable. Even where that is the belief, there is usually a tradition of at least being aware of the presence of god and even of receiving communications from god. How does that work? I have proposed a hierarchy of intelligences and said that we can perceive those entities a few levels above our own. If the god that people can perceive is just the highest level of the hierarchy that humans are capable of perceiving, then we can accommodate both ideas: It is possible to communicate with god, the one we are capable of perceiving, and god is unknowable, being the levels above our ability to perceive. The two are just different views of the hierarchy.

Once again, I have slipped into the language of isolated material beings. I have been using phrases like "hierarchies of entities" to describe thought/feeling in the field. That makes it sound like a bureaucracy. The worker reports to the group leader, who reports to the department manager, who reports to the area manager, etc., up to the CEO who lives on the top floor. And these are all separate people who communicate up and down the hierarchy as they choose, or not.

The field does not work this way. All the waves are in the same "place." Connection, interference and influence are instantaneous and continuous based on resonance of the waves. All of the entities I have been referring to are just aspects of the single quantum field.

There is even a bit of old wisdom to describe that: All Is One. The "One" part does not refer to physical things, which are obviously not anything like "one." It refers to the spiritual realm, which I have identified as the thinking/feeling quantum field. There everything is, indeed, one cohesive field of thought/feeling, influence and connection.

The nature of god

If that is what god *is*, then what is that sort of god like? God is described in many ways today. In the West (I think I can include Christianity, Islam and Judaism in Western religions), god is viewed as both all good and all-powerful. In the East everything, including god, is viewed as a dynamic balance of opposites, yin and yang. There are many variations in both categories. The quantum-field-as-god model makes some predictions about the nature of god.

- The form of the god perceived by different individuals will vary with the individual and with his/her cultural background. Because the field is formless, people can wrap any form they choose on the being they communicate with. This is our normal perception process.

- Because the quantum god is part of (or is) the intelligent field, it is possible to feel the presence of that intelligence and to communicate with it through our intuition. Like all intuitive insights, the information received can be wrapped in any sensory input we choose.

- The quantum god exerts continuous influence over the material world at the quantum level. This means the influence of god is very subtle and very profound. Humans can perceive this influence because of their innate abilities in statistical analysis, that is, the ability to see subtle patterns and trends in their environment.

- The influence that the quantum god exerts is exactly the same as the influence that humans exert through the quantum field. The only difference is in the scope or scale of the influence, which precisely matches the respective positions on the scale of awareness.

- In addition to being subtle, the influence of the quantum god is probabilistic, not deterministic. The quantum waves that are the medium of influence do not define specific outcomes. They define probability curves for the outcomes of many individual events.

- The mechanisms I have described do not explicitly account for good, love or any other value sorts of things. The model does say that the field runs on resonance and harmonics. The interaction between harmonic waves is the basis for all communication and influence. In the coming chapters, I will use this idea to account for the power of love and good in our universe.

I think these six items are quite consistent with the observations. They are decidedly not consistent with many of the specific views held by practitioners of various religions, but then I think many of those views are not very consistent with observations. The idea of a god or gods is widespread and persistent, but the specific image of god varies widely over individuals, cultures and history. That is what you might expect if people are wrapping their own interpretations on a formless entity.

The same is true for communication with god. People are very definite and even defensive about the identity of the entities they communicate with. That people report a variety of communication forms lends support to the idea that they are wrapping their own perception on a more abstract stream of information.

In some places, god is described as an absolute, all-powerful force for good. The implication is that the force is deterministic, which is what we usually think of when we think of "powerful" forces. As I look at the world, I see a pretty even distribution of trouble and tragedy. Nobody seems to have a corner on goodness. People with money are able to avoid many problems that are common in people with no money, but that seems to be directly related to having money and does not appear to have much to do with spiritual state or membership in any religion. The Eastern religions describe god, or the universe, as a dynamic balance between opposites: yin and yang. This description is closer to the observed world. The world is a mix of outcomes, from triumph to tragedy, good to evil. We can describe those outcomes with probability curves, but does that imply any intelligent influence, or is it just a random distribution?

We can't tell just by looking at the outcomes of events. To see the influence of intelligent forces, we need to rely on human perception. People have always detected the effects of influence on outcomes with the innate human ability to do statistical analysis, that is the ability to detect patterns and trends in mixed outcomes. There are also many studies of the ability of people to influence outcomes of both human and natural events (see Chapter 10). The influence that is observed is not an absolute shift of all outcomes. It is a change in the shape of the probability curves so that there is more of the desired outcome than there was before, but there are still plenty of the undesired outcomes.

In the next section, I will talk about some situations where the scope of influence is huge, such as the big bang and the evolution of species. These are excellent examples of what I mean by large scale, subtle and profound influence. They are well beyond what I would expect from human-scale awareness.

So there is something bigger "out there" in the world of quantum intelligences. When the scale of the intelligence gets big enough that we can just barely perceive it, we call it "god." Next I would like to examine the world around us for hints of what that large-scale awareness and influence might do.

Evidence for active connection

Everything seems to work at all scales, from quarks and atoms up to the entire cosmos and the way things seem to work tells us that there is non-local influence at all those scales. Well, maybe that's a little strong. It obviously does not tell everyone that there are non-local influences. None of the explanations in current science involve non-local influences. So what is it that tells us that non-local influences are present? The short answer is, "Explanations that involve one or more steps that are extremely unlikely."

The Big Bang

The big bang is a good example of an explanation that involves highly unlikely steps. How unlikely? Roger Penrose calculated[1] that the likelihood of our specific universe emerging from the big bang event is one part in ten raised to a number that is 10^{123}. That makes winning the Powerball jackpot look like a sure thing. For comparison, the number of atomic particles, protons, neutrons and electrons in the visible universe is 10^{73}. The visible universe is pretty big: 30 billion light years in diameter, with lots of very heavy stars and planets.

Life and Spirit in the Quantum Field

Protons and neutrons are very small, so there are a great many of them in the universe, but I can write that number as 10^{73}. The odds against our universe emerging from the big bang are vastly larger, one part in ten to the ten to the 123. We are a very unlikely event, but we are definitely here.

Evolution

Another scientific explanation that involves unlikely events is evolution. The current explanation of how evolution happens is random genetic mutations coupled with natural selection. What is the likelihood that a single genetic mutation in a single individual would spread to the huge number of individuals that make up the species? It seems very unlikely.

There is another unlikely step in the current model of evolution: A sequence of random mutations in individual genes can produce smooth, coherent change in the form of a species, as in giraffes evolving from sort-necked animals to long-necked animals. The evolution of giraffes, growing from an animal with a short neck to an animal with a very long neck, requires changes in many genes. The likelihood of "random mutations" producing that coherent sequence of changes is exceptionally small.

Snowflakes

Large events are not the only phenomena that have unlikely explanations. Consider something small, like a snowflake. Many snowflakes have a beautiful, six-sided symmetry. Snow crystals can take many forms: planar crystal (the pretty ones), tubes, columns and needles,

Chapter 14 Mechanisms of Spirit

depending on the temperature and humidity of the air where the flakes are forming. Snowflakes form in the air when water vapor condenses directly into solid ice. They consist of around 10^{18} molecules of water. That's a lot of molecules, but all together they weigh only three hundredths of a milligram, 0.03 mg. The flake grows from a small nucleus, outward. The unlikely event in the growth of a snowflake is that molecules of water condensing into ice at the end of one of the arms of the snowflake would form the same shape of branches that are forming at the ends of the other arms. At the molecular scale, the ends of the branches are very far removed from each other. If condensation of a single molecule is a local, isolated event, then it is very hard to explain why all those individual condensation events at the ends of those widely separated six arms would just happen to produce the same pattern of branches on each arm.

The problem of large-scale influence

I am not unique in worrying about unlikely events in scientific explanations. Rupert Sheldrake, the British biologist, has been concerned with these sorts of problems for a long time. He has proposed a thoroughly outside-the-mainstream-paradigm solution to the problem of unlikely outcomes, which he calls morphic fields. These fields direct the development of all material and information forms. They act across all space. The connection between a particular morphic field and the specific instance of the thing it guides is made via a resonance between the field and the thing. In his book, *The Presence of the Past*[2], he talks about fields in general, specific fields known to science and his morphic fields. He mentions in passing that it might be possible that morphic fields are the same as the quantum fields, but that the mechanism of the influence remains unknown.

The position I have taken in this book is that the quantum field provides the functionality that Sheldrake describes for morphic fields.

The explanations given for the big bang, evolution, and even for snowflake formation are unlikely because of our assumption that all the individual pieces are acting independently of one another. All of those independent events make the odds of the observed outcomes very small. A good way to make the events more likely is to say that the pieces, atoms, genes and water molecules are not independent. They are connected at the quantum level to a coherent, intelligent, causative field. The calculation of the likelihood of the observed outcomes would be very different if all the pieces were entangled in the quantum field.

Saying that there is this field coupling does not explain why we emerged from the big bang or why giraffes got long necks. Let's consider intelligence next.

Influence and who is doing the influencing

If all these things, from the big bang to the formation of snowflakes, have these unlikely explanations that can be better explained with non-local influence, what is the nature of the influence and "who" is doing the influencing? The nature of the influence is easier, so let's start with that.

In our regular, material world, we make a distinction between "looking at" and "moving." We can observe things without influencing them. When we want to influence something, we have to exert some sort of force. That's Newton's first law of motion: An object remains at rest or in uniform motion until acted on by an external force.

Things are different at the quantum level. We saw that just observing, or even being able to observe, the outcome of a quantum level event, like single photons going through two slits, is enough to alter the physical outcome. We saw in the PEAR machines that intent to alter the pattern of random material events could change the pattern of those physical events. "Intent" has more of the feel of a force than a "just looking" event, although both are thoroughly non-material. We also saw that the PEAR machines were influenced by coherence of thought or focus of many people. That is, people can exert an influence over a physical, though quantum-level, event without intending to do anything.

I would like to generalize these observations by suggesting that "connection" of any sort *is* influence. That is, the simple act of connection influences both connectees. There is no force needed, where force is something different from just looking. In the field connection is ubiquitous and connection is the same as influence.

I gave some examples of situations that called for action at a distance, ranging from the big bang down to evolution of species and on down to the formation of symmetric snowflakes. There are no human minds thinking in any of those situations, so how are the connections made? The same way they are made when human minds are involved: by resonance. I don't think that the presence of human intelligence or consciousness is necessary or even important in making these connections in the quantum field. Recall the example of holographic computer memory. The only thing required to retrieve a piece of information (that is, to connect with the information) was to put out a tiny fragment of the information wanted. Even in the crude state of the present research in holographic memory, only 0.05% of the original information was needed.

The information part of the evolving giraffe, or the growing snowflake is distributed in the quantum field. The simple act of putting that information in the quantum field allows it to resonate and interfere with all similar information. That interference shows up as the orderly evolution of all giraffes and in the symmetry of snowflakes.

This connection is called entanglement when we are talking about photons and electrons. The suggestion I am making here is that the phenomenon of entanglement extends to higher level entities, like snowflakes and the genome of giraffes. We will see in a few pages that entanglement involves more than the physical properties of electrons. There is intelligence involved.

It's the other way around

Well, I did it again. I said, "putting that piece of information in the quantum field." Old habits die hard. That makes it sound like the information originated in the material world so it could be "put" into the quantum world. It is really the other way around. The information, the resonance and the interference all originate in the quantum field. The results of all that interaction get projected into our material world via perception or interactions between material objects. Even using the word "connection" to describe what is happening is really a remnant of Newton's old view. "Connection" implies that there are two separate things that were isolated from one another and they somehow gained an attachment with each other by becoming connected. If all movement originates in the quantum field, then the state of "connection" is the normal condition. Everything is connected to everything else that has any sort of common resonance. Things only become "disconnected" when they are projected onto the material world.

Nature of the influence

I have even been describing influence in the same way I would describe Newton's force acting on a material body in uniform motion. The body is sailing along, minding its own business, when, suddenly, wham! some outside force bangs into it and makes it change direction. So I have implied that the pre-giraffe is minding its own business, eating grass with its short neck when suddenly, wham! a wave of influence from the quantum field hits it and changes its genes so its children will have longer necks. That's not the way it works. The influence, as I have called it, is the continuous interaction of resonating fields in the quantum field. These fields define the probability curves for the outcomes of all quantum level events in our material world. In the case of our pre-giraffe, the interaction of the fields moves the probability curves for giraffes toward having longer necks. Then all of the events controlled by those curves, which include the conception and development of all giraffes, tend to produce more giraffes with longer necks.

Life and Spirit in the Quantum Field

While giraffes are not quantum level objects, the conception and development of giraffes is a quantum process. The probability curves of the field govern the collective outcomes of all those individual quantum events, that is, the development and birth of individual giraffes. This is possible because the quantum processes of all giraffes are entangled. The "influence" that worked on pre-giraffes moved those curves so that, on average, the necks got longer. All of the needed systems change at once. The development of an individual giraffe is not an isolated event. It is not separated from all the other individual giraffes.

I can use the same sort of reasoning to describe any of the phenomena that suffer from unlikely steps in their current explanations. All of the components of the quantum soup of the big bang were entangled, by definition, since they all originated from the same point. Their individual actions were not isolated from one another. This means that the assumption of isolated particles in Penrose's calculation is not correct. The individual quantum events that made up the history of each particle were (are) guided by probability curves in the quantum field. The probability curve applied to all the particles. The outcomes of the individual events are not independent of each other. They are guided by probability waves at the scale of the entire universe.

From the cosmic to the minute, consider snowflakes. The unlikely step is the condensation of isolated water molecules in six-sided symmetry across the vast distances (at the molecular scale) of the growing snowflake. If the individual water molecules become entangled by joining the snowflake, then their condensation patterns can be governed by probability waves in the quantum field. The condensation events of the individual molecules are all covered by the probability curve for that snowflake. The growth of the snowflake is not a collection of isolated condensation events, it is the expression of a single set of probability curves. The process is exactly analogous to the formation of the interference bands in the double slit experiment. Even when the photons were being sent through the slits one at a time, they were still governed by the probability curves for that experimental setup. Individually, they appear to be random events, but collectively, they came out exactly as the probability curves suggested.

Now that we have all these individual events connected, let's consider why the snowflake comes out pretty and why pre-giraffes evolve into giraffes that are whole, functional animals.

The quantum god

This sort of quantum level influence is what the quantum god does. It is subtle, but profound, shaping everything from the form of the universe to the form of snowflakes.

In order for the quantum-level connections I have described to make a snowflake symmetrical or make giraffes get longer necks, there must be information and intelligence in the connection. As I have described it, the connection imposes the form and structure on the individual parts. In a snowflake, the parts are water molecules. In a giraffe the parts are genes, cells, tissues, organs systems and organisms. There must be a great deal of intelligence and information. Where does that come from?

Who is doing the influencing?

Back when we started talking about influence, it was described as the force. If it is force in the conventional sense, then it has to have a source. Someone or something has to "exert" the force. When a healer is healing a patient or people are praying for someone's recovery from disease, it is easy to see that the healer and the praying people are the source of the influence. But since then we have talked about influence as being a harmony of waves in the quantum field, where the waves are the probability curves for the outcomes of events. Harmonizing with the waves is the same as changing the outcomes, which is what we originally called influence or force. The force is not a separate thing, it is the guide for the outcomes. The waves of influence *are* the waves of intelligence. The force *is* the intelligence at any scale.

When I introduced the analogy of singing in Chapter 13, I necessarily introduced the idea of a "singer." Who is making the harmonizing vibrations? Once again, I think the Vedas can help us here. In the Vedic traditions, Om is the primal sound and the primal creative force. It is also the name of god, the one who made the sound. Is the same word "Om" being used to describe two different things: a sound and a god?

The answer is yes. The sound and the god are the same thing. If this sounds weird, consider the same idea expressed in the Western tradition. "In the beginning was the Word, and the Word was with God, and the Word was God," John 1:1. The difference between "Word" as used in John and Om in the Vedic scriptures is very small. The small difference gets even smaller when John is read in the original Greek, where "logos" is used. Logos is much closer to the idea of Om as the basis of all creation.

Back in the chapter on biology (Chapter 7), I described the form of our thoughts and feelings as being the same form as the quantum field, and I concluded that the quantum field thinks/feels with the waves. The waves are thought/feeling. But those waves are also the "force" that guides the outcomes of material events. So we are back to the Bible and the Vedas. Om and God are the same thing, Logos and God are the same thing. Let me put this back in secular terms: the influencer, the influence and the outcomes are all the same thing – the living hologram that is the quantum field.

Who does the influencing? The answer is that everything influences everything that he/she/it has any thought or feeling about. And by everything, I mean all the waves associated with material bodies and all the waves that are not associated with material bodies.

The quantum god is the form and the connection that allows cosmic numbers of apparently separated pieces behave in a coherent and creative way. It can do this because it is both intelligence and probability waves for all quantum level events. Another way to say that is that god is the ground of all being. This was humanity's first conceptualization of god. It is still active today in many mystic and indigenous belief systems. I am happy to see it re-emerge from thinking about quantum life.

God and spirit are real in scientific terms, although it is still fringe science. In the next chapter we'll see if we can learn anything about spiritual values from our new model.

[1] Penrose, R. *The Road to Reality*, Alfred Knopf, New York, 2004. Pg 726-31.

[2] Sheldrake, R. *The Presence of the Past*, Vintage Books, New York, 1989.

Chapter 15
Spiritual Values

SPIRITUAL TRADITIONS AND practices all over the world are concerned with values, morals and right living. What is the best way to live our lives in this world and any others we may visit? I have been using the language of secular and value-neutral science to get this far. It's not that I expect that there will ever be a equation whose solution is, "Do unto others . . . ," but I wanted to see if I could find any support for the value of goodness in the quantum life model.

Values as connection

The path I found to value was through connection. The quantum field is resonance and connection. It is so connected that there aren't any things left to be connected. Only the connections remain. Let's see how that relates to spiritual values.

Spectrum of feeling

Spiritual teachers throughout history have placed great importance on love and compassion. Unfortunately, many of the organizations that bear the names of those teachers don't seem to value love and compassion as much as their founders did. Nonetheless, it remains an important part of spiritual teachings. Using my innate abilities in statistical analysis (the same abilities that everyone has), it appears to me that there is more value in approaching people and situations with love and compassion than with, say, physical force or anger.

Love and compassion are half of a spectrum, the other half of which is forgiveness and acceptance. The complete spectrum is forgiveness, acceptance, compassion and love. Behaviors based on that spectrum define what we mean by goodness. Love does not show up in the model any more than goodness does, but I believe we can account for the power of those attributes if we view them as degrees of connection. You might feel differently toward someone you have a need to forgive compared to someone who inspires love, but I believe the difference is more of degree than of substance. They all represent a connection between two people and the connections all represent varying degrees of positive relationship.

High level behavior?

We like to think our high-level human traits are unique to humans. The lower animals can't do what we do. Humans have emotions like love and compassion. We solve problems. We court each other. We understand physics (At least some humans understand physics). I think that it is really the other way around. We are animals and do mostly the same things that other animals do. The difference in humans is that we talk about it and declare some behaviors not appropriate for polite society. Consider these examples. Have you ever watched a squirrel trying to get into a squirrel-proof bird feeder? Or seen courtship in animals or birds?

Chapter 15　　　　　　　　　　　　　　Spiritual Values

Richard Feynman, the Nobel physicist, had an argument with another physicist about whether dogs understood physics. Feynman said they did. The other physicist said they didn't. Feynman carried the argument when he asked, "Have you ever watched a dog catch a Frisbee?"

Why should our behavior be the same as the lower animals? We saw the reason in Chapter 10. Our behavior is run by our feelings and emotions, not our uniquely human rational thought. Our feelings and emotions are carried in our bodies by the neurotransmitters and we share the same neurotransmitters with every other member of the animal kingdom. The basic behaviors of all animals are driven by the same emotions. Each species expresses the behaviors differently, but the behaviors are basically the same. Dogs and worms can be in love, it's just that they can't write love songs and play them on their iPods.

Much of what we think of as unique human behavior is just human-level expression of behaviors common to all animals. I would like to carry that idea one step further and suggest our highest level behaviors, forgiveness, acceptance, compassion and love, are human-level expressions of something more fundamental than shared neurotransmitters: connection through the quantum field. Specifically, I suggest that love is awareness of that non-material connection between people.

To make connections in the field, we need some common tag or fragment, some harmonizing vibration. We provide that vibration when we forgive, accept, are compassionate with or love another person. The field is connection. It has been called a dance with no dancers. There is just the dance. The dance is connection. When we make that kind of connection it feels good. Connection in the field is the same as influence, so we are most powerful (in a subtle but profound way, of course) when we are connected with people.

While I did not solve any equations to arrive at "love one another," this rationale linking connections in the field to our highest feelings encouraged me to think that I might be on the right track.

The other side

I have painted a picture where love and compassion are the ruling forces, but you don't have to look very far in any direction to see that there are other forces controlling much of human behavior. People treat other people without compassion and lots of people treat themselves without compassion. So, the bad stuff works. You can make bad things happen and people do it "successfully" all the time. This is somewhat troubling because I would really like there to be a higher value on good than bad. Clearly, there is not a large advantage to being good, since the bad stuff seems to be holding its own, but I would feel better if there was at least a little advantage for being good over being evil.

The clue that the universe values good over evil comes from healing practice. Healing, as we will see in the next section, requires acceptance, and acceptance is connection. You might be able to build an empire or a corporation with evil acts and attitudes, but you can't heal yourself or anyone else with that attitude. So I happily conclude that the universe places a slightly higher value on acceptance, compassion and love than it does on anger, envy and the rest.

Chapter 15 — Spiritual Values

Acceptance as power

Healing means to help another move toward wholeness, which may or may not include reducing the negative symptoms. I have noticed that there is a difference in the effectiveness of healers. I have seen this sort of difference in doctors, dentists, nurses, bar tenders, and barbers. You probably know of well-qualified and duly licensed doctors that you would not consider going to. They have a bad reputation or bad vibes or something that makes you stay away. You probably also know doctors that are a joy to visit. Just being in their presence makes you feel better.

I have wondered what caused those differences. When I was taking my energy healing classes, a common message was that it is not the healer who is doing the healing it is the energy that flows through the healer. This is good advice because it helps keep the healer's ego in check. Strong egos send bad vibes when they are waving their hands over you in any healing technique. Now that I think about it, hand waving is not required. One can feel bad just sitting in the office of an egotistical medical specialist.

In spite of that advice, I also noticed that some healers felt much better and were more effective than others, so the healer clearly made a difference. Some people feel great to be around and some feel bad. This appears to be independent of the skill level or qualifications of the person.

I found the reason for the differences in a story by Rachel Remen, M.D. in her book, *Kitchen Table Wisdom*[1]. The story described a lecture and demonstration by Carl Rogers that she attended in the late '60s while she was on the faculty of Stanford Medical School. Rogers was a psychologist who developed patient-centered therapy in the late '40s. The assumption in his model was that the patient knows what is needed for his or her healing and the therapist's job is simply to help the patient discover it. By the time Rogers gave that lecture, he had reduced the technique to its essence. He called it Unconditional Positive Regard.

Carl Rogers and healing

After he described the technique, Rogers invited one of the doctors in the audience to be the client for a demonstration They sat on the small stage facing one another. Before he started the demonstration Rogers said this to the audience. "Before every session I take a moment to remember my humanity. There is no experience that the man has that I cannot share with him, no fear that I cannot understand, no suffering that I cannot care about, because I too am human. No matter how deep his wound, he does not need to be ashamed in front of me. I too am vulnerable. And because of this, I am enough. Whatever his story, he no longer needs to be alone with it. This is what will allow his healing to begin."

Rogers simply sat there in a state of deep acceptance. The doctor began to talk. When he finished talking he, and most of the audience, had had a profound healing experience. There was no outer technique involved. The healing took place because of Rogers' inner state of complete acceptance of himself and his patient. Rachael Remen summed it up nicely when she said, "What Rogers was pointing to is, of course, a very wise and basic principle of healing relationship. Whatever the expertise we have acquired, the greatest gift we bring to anyone who is suffering is our wholeness."

I looked up some of Rogers' books from the late '40s where he described how he trained people to do this kind work. He said the technique rests on acceptance of the patient by the therapist, and to accept the patient, the therapist must accept him/herself. The catch is that to accept yourself, you have to get past your ego. Rogers' writings seem to have been influential, but his techniques were never widely adopted. I can understand why. How many PhD psychologists or psychiatrists do you know who would be willing to give up their egos?

Rogers was able to heal people using no physical technique. He "just sat there" in a state of deep acceptance. When I read that, I found the answer to why healers vary in effectiveness. Acceptance is the ingredient that determines the effectiveness of all healers. And, acceptance, as we saw before, is how humans interpret connection at the quantum level.

If you are aware of and accept yourself, you are whole. If you are whole, you are capable of healing others because you can accept others. Being able to accept self and others, you connect with all. You can connect with the spirit, the non-material intelligences, and that makes you holy. Being healed is to be whole, being whole is to be holy.

All this acceptance, connection and healing is working through the quantum field. The connection is influence, but the influence works at the quantum level on quantum level processes, like human health. Rogers got "amazing results," perhaps better than 99% of therapists, but he didn't heal everyone. Actually, even a

50% healing rate would be pretty amazing in psychotherapy. The best influence we, or anything else, can exert is to move the probability curves. We cannot make the curve into a single point.

Ye are gods

Ye are gods. This bit of old wisdom is from Psalms 82:6. It sounds pretty outrageous today. In most versions, god is all good, all knowing and all powerful. I don't know any humans like that, so how can we be gods? We have seen that humans are more powerful than our cultural norms would allow and god is less like Zeus than some religions would like to believe, so it might not be entirely outrageous.

The possibilities and limitations of human divinity are nicely illustrated in a 2003 movie called Bruce Almighty. Jim Carrey is Bruce and Morgan Freeman is God. Bruce lives in Buffalo, NY. God calls Bruce to an empty office building. God turns out to be a janitor mopping the floor. Dressed in white, of course. God asks Bruce if he wants to help mop the floor and Bruce declines. God tells Bruce that he, Bruce, is going to become God. After a bit, Bruce gets into his new powers, but troubles follow. Bruce is overwhelmed with the number of petitions for help and his power doesn't help get his girlfriend back. Things go from bad to worse and soon there is rioting in the streets. Bruce goes back to God for help. Bruce finds that the way to restore order is to mop the floor: a subtle but profound approach to power.

Bruce says that he can't handle the problems of the whole world. God says that he didn't have the whole world. He only had a few blocks of Buffalo: a nice example of the scope of awareness and influence that we talked about.

How can we humans be gods? God is connection and influence and living information in the quantum field which makes for a very subtle god. We humans (and all other living things) are also connection and influence and living information in the quantum field, just like god, but with a smaller scope of awareness. That makes humans powerful, but still subtle. The main difference between humans and god is the scope of awareness each has. The kind of knowing and influence is the same at all scales. Humans, of course, also have material bodies some of the time.

The old bit of wisdom turns out to be true. To see the truth of that idea, we have had to invest humans with god-like powers and make god subtle and probabilistic, but it was worth it. If people are gods, maybe we will be nicer to them. Maybe we will be nicer to ourselves. We have the power.

God and spirit in Newton's world

We have arrived. We have seen all the pieces: the physics of the quantum field, the quantum processes of mind and body, the workings of spirit, soul and god in the living quantum field. What can we say about our current world? I think we have to say that the separations are not serving us. The distinctions between material science and transcendent religion, spiritual and material, body and mind, people like us and people not like us are not justifiable. Those distinctions cause us vast pain and suffering. But we already knew that.

Life and Spirit in the Quantum Field

The new/old information is that god and spirit are not "out there" someplace far away. They are the ground of our reality. They are here. They are us. The secular, mundane world we live in is a projection not just of the quantum field, but of god.

Our modern, scientific culture started off by separating the material world from the transcendent. The material world was called real and everything else was not real. The world was a clockwork construct following immutable laws of nature. Meanwhile, the transcendent world lived on in the hearts and minds of people everywhere and in lots of organizations, some of them large and powerful. Measurement and observation ruled in the clockwork world. They were not important in the transcendent. The clockwork world turned out to be made of little balls bouncing around and bumping into each other.

For all their differences, the two worlds shared many features. Both worked on paradigms and those who spoke heresies against the local paradigm were treated very badly. The material world emphasized isolation of material bodies. The transcendent world emphasized the isolation of spiritual bodies.

A hundred years ago, there was a paradigm change in the physics part of the clockwork world. In the new paradigm, the littlest balls were not isolated. They were entangled, across time and space. All (of the little stuff) was one (at least until it hit something). Cracks appeared in the material world and transcendent light was shining through.

Thirty years ago, that new paradigm started finding its way into the life sciences. Now, it turns out that living things have eternal parts. Living things are both material and transcendent. The transcendent parts are all one with all the other transcendent parts. The world that the clockwork people called spiritual and not real turns out to be real. It is not separate from material reality, it is the ground of all being. In particular, it is the ground of all living, thinking, feeling being. It is the ground of us.

This new quantum life paradigm is a big change for the clockwork paradigms in physical and life sciences. It is also a big change for many old paradigms in religion. God, soul and spirit have all become part of the real (scientific) world, but at a price. God may still know everything, but the power of god has been reduced to moving probability curves for quantum level events. The scale of the power is still awesome and there is still much about god that is completely beyond our ability to perceive. So, god is still ineffable.

The result of all this is that the isolation that characterized scientific reality and much of religious reality has been replaced with a reality where the massive, material stuff is still isolated, but all the little material stuff, and all of life exists in a single cloud. Everything is the entire cloud. The cloud is full of resonance and harmonies. The cloud is singing to itself and the sounds animate the universe.

How far have we come?

This time we have taken some more giant steps. We saw that Descartes' transcendent domain is alive, intelligent and influencing our material world all the time. Our non-local, non-temporal feelings are in heaven all the time, whether we have a body or not. That is the nature of things distributed in the field.

The field is intelligence, feeling and action. On one scale, it animates bugs and mice, on another it animates us, and on a large scale, we call it god. The behavior of the universe at all levels is coherent across time and space. A quantum intelligence at every level can account for that coherence.

Resonance is the mechanism of action in the field. We saw that connection by resonance is the basis for the traditional values of acceptance and compassion. Acceptance is how we express power through the field.

If this is how the world works, what does it mean for the way we do things? Given where most of Western culture is starting from – Newton's clockworks with Descartes' separation of transcendent from material – the answer is that big changes are needed to live the possibilities offered by the new reality. We will look at some of those changes in the next part.

---

[1] Remen, R. *Kitchen Table Wisdom*, Riverhead Books, New York, 1996. Pg 217-19.

Part 5 Living in the New Reality

How we live depends on how we think the universe works. We have spent the previous three parts of this book developing a model of how things work. What would it be like to live in a world where most people subscribed to that model? It would be different than the way we live today. Jobs and institutions would be different. Individuals would behave differently. They would also be practicing and developing different skills.

In this part of our journey, we will look at how things might be different if our model moved from the fringe to the mainstream of our culture.

The quantum life model describes abilities that all people have, but that few have exercised. If you wanted to use those abilities you would have to learn some new skills, which would take some practice. A serious discussion of practice is beyond the scope of this book, but I did put a few words about it in the appendix.

In this part we will look at what life might be like if people embraced the quantum life model and its possibilities.

Chapter 16
The Sciences

THE QUANTUM LIFE paradigm we have developed provides mechanisms of science that allow us to be very different from what was allowed in Newton's paradigms. It allows us to be whole and holy, to be connected to all things and to be influential in all things at all times, and even outside of time. I would like to consider how this might play out in the sciences, starting with the physical sciences and then moving on to medicine and healing.

Let's begin with science.

The implications for science as we know it today

The big change for science as we know it today is this: The experimenter is a force that influences the results. The experimenters' expectations at the felt-sense-level exert forces on the quantum-level outcomes of events in the experiments. What does this mean for science?

Objective reality

Scientists like to believe that they are observing a reality that is separate from themselves. This has never been true. Even without the quantum mind ideas we are talking about here, it was recognized (not widely, of course) that the goals and intentions of the scientists making the observations affect what they see.

Now we have an even stronger connection between the observers and the observed. First, the personal beliefs of the scientist affect functions at all levels of perception: from low-level sensor processing in the brain up to human-level interpretations of meanings. Next, we have human thought and intention influencing what happens in the "physical" world at the quantum level. Some outcomes are more susceptible to quantum-level effects than others. The outcomes in nuclear physics are all quantum-level events, so are the outcomes in human health (see the next section on implications for medicine). The outcomes for moon rocket trajectories are not very dependent on the quantum level since most rockets are very massive.

Science and scientists run on paradigms, just like people in every other domain of life. And just like everyone else, scientists take their current paradigms to be reality, usually for their whole careers. That changed in nuclear physics and cosmology as a result of the quantum and relativity revolutions. The old reality of the 19th century changed so quickly that everyone involved was forced to acknowledge that the paradigms are not reality. They are models and the models can change very quickly with new discoveries. The result was a rule for new paradigms. It states that new paradigms have to produce the old rules as special cases. That is why quantum mechanics reproduces Newtonian mechanics at high mass. Relativity produces Newton's laws at low velocities and low gravities.

The result of all this is that objective reality is not what it used to be. We already knew that whatever is really

Chapter 16   The Sciences

out there is filtered through our paradigms. Now we know that scientists can influence the results in subtle but profound ways. We are not separate from what we observe.

Training scientists

Training scientists has always been a process of indoctrinating students into the current paradigms[1]. Most scientists then take their paradigms to be reality. One change that I might expect to to place a greater emphasis on the transient nature of paradigms.

Another change is that the scientists' inner expectations are a force that influences the outcomes of experiments. Actually, it is worse than just the direct experimenters having an influence. The influence is exerted across time and space, so the list of people exerting an influence on an experiment can be large. Anyone, anywhere, who knows about the experiment and has feelings about the outcomes, exerts influence. So the sponsors of the research, other researchers in the field (both those who support the theory being tested and those who support competing theories) and people who want to use the outcomes, all exert subtle influence on the results. The list of influencers might include people like designers waiting for an improved plastic to make their widgets last longer and patients waiting for an improved drug to treat their disease.

This means that scientists will need to be trained to be aware of their own inner feelings and expectations and the inner feelings and expectations of those who have any interest at all in the outcomes of their work. This is a big change for a field that attracts people who, in

295

general, do not want to deal with inner feelings and expectations, their own or anyone else's.

Cause and effect

The idea that effects have a physical cause is fundamental to Western science and to lots of non-science people. It is hard to imagine a world where things can happen without a physical cause, but the lockstep relationship of cause and effect has already been challenged a couple of times in the last century.

Relativity told us that an observer moving fast relative to us might see the outcome before he sees the cause. Quantum mechanics told us that the outcome of a measurement made in the past can be influenced by observations made in the present. Now we have the idea that the outcomes of experiments can be influenced by a non-physical "force." These entirely non-physical forces can cause some physical outcomes. This might be frightening if you require a physical cause for all your results, but it is not too bad if you can allow for non-material forces. It's not that there are no causes in this new model, it is that the forces that influence outcomes include those of intention and expectation acting at the quantum level. William Tiller has demonstrated the explicit effects of intention on the outcomes of physical measurements.[2]

How experiments are done

Experiments are set up very carefully to control all of the influences that might affect the outcome. The goal is to keep all the influences but one constant and vary just that one. We have added some more influences that affect the outcomes: the intentions of all the people who are interested in the results. So, how do we control those influences?

Two possibilities come to mind. One is to have all scientists be trained as Buddhist monks so they can effectively control their inner state and be aware of what it is. This would help with the experimenters, but would not help with the managers of the scientists or with the people wanting to use the results. I don't think this is a very realistic suggestion. Another more realistic approach is to use statistics. When all the variables cannot be controlled, the approach is to do many experiments and take averages. If the averages are carefully done, it is possible to tease out the effects of individual variables.

Consequently, we might have different groups of experimenters do the same experiment: one group known to expect one outcome and another group known to expect another outcome. The problem is complicated by the non-locality. It is possible to imagine controlling who actually works on the experiment, but it seems very difficult to control the expectations of all the people who are interested, but not directly involved. That might include people like the management of the company and competing companies and people who might benefit from one kind of outcome or another.

Science and spirit

This discussion is often described as being about science and religion, usually referring to organizations of science and organizations of religion. This is a very difficult discussion because each religious organization is necessarily trying very hard to maintain its position of holding the unique truth about spiritual matters. Science organizations are trying very hard to maintain their positions of holding the unique truth about material reality, which, by definition, does not include non-material spiritual stuff.

I would like to cast my discussion on this topic as being about abstract Cartesian science and abstract belief in a spiritual reality. The wall between the two was thinned considerably when quantum mechanics was developed. People in science have been talking about that for 80 years. The proposals made in these pages make the thin wall a very thin veil. The spiritual implications of Schrodinger's wave equation are just the tip of the iceberg. We have seen that in addition to photons and electrons, the quantum field is soul, spirit, spiritual beings of all sorts and god. Science has undone Descartes' deal to divide reality into the material and the transcendent. The transcendent turns out to be not so far removed from the material. Some scales of beings in the field are certainly unknowable, but many of the levels are accessible to us. We experience interactions with the field at all levels of our being.

Science has opened the door to spirit and future developments, such as the proposals in this book, will force a recognition in science that there is more to the universe than matter and physical forces. Perhaps "force" is too strong. The recognition will more likely come as the current generation of scientists is replaced with the next.

I don't think that the recognition will take the form of any official reconciliation between scientific and religious organizations. Rather, I think the quantum revolution will continue to creep into mainstream science. If the creep follows any of the suggestions made in this book, eventually science will uncover intelligence in the field. When that happens science will be fully in the spirit business, regardless of what they call it.

I am hopeful that the same sort of process might work its way through organized religions. At some point, religions might discover that the same set of phenomena underlie all religions. At that point, reconciliation of science and religion would be an easy task. Science and religion do not need to be the same thing. It will be nice when they are not trying to be mutually exclusive defenders of the true reality.

I think the anti-spirit parts of science are destined for change. It may take a generation or two, but the change is underway. The change will necessarily increase the importance of the observer's inner state. The inner state of the observer affects the interpretations of the results, influences the outcomes and can be used as way to obtain information about the experiment (via intuitive knowing). This will make science look a great deal like the work of the early mystics, exploring the inner spiritual landscape with their own powers of observation and intuition.

Now that I look at it, that kind of change will probably take more than one new generation of scientists to make it up to the mainstream.

The implications for medicine as we know it today

Applying this new model to medicine is very interesting. On one hand, medicine is the most defensive science. Western medicine has a long history, and current practice, of suppressing ideas and therapies that do not fit in the current paradigm. On the other hand, the subjects of medicine, people and their health are very susceptible to the non-material influences of human intentions and expectation. Human health is a quantum-level phenomenon that is driven by quantum-level forces coming in part from human minds and feelings. Health and behavior are influenced by collective intelligences. This is a big change.

Medicine is, of course, facing lots of other challenges. Treating people like machines is beginning to backfire on the HMO planners and economists. People don't like to have 12-minute sessions with their doctors, so they are going to complementary/alternative practitioners, many of whom take into account peoples' intentions and expectations. I think change is under way, and the direction of change is toward recognizing more than material influences on health.

Ken Wilbur has proposed "integral medicine"[3] where influences from both individual and collectives are included in the care process. The model I have described here provides a mechanism for how those influences work.

Let's consider some specific areas that are impacted by the quantum life model.

The model of health, disease and health care

Right now, the model of health and disease is that disease is caused by material factors: external microbes or toxic substances entering the body, or a material breakdown of one or more organs in the body. Treatment involves addressing the material failures inside the body, eliminating the external microbes or substances, and

Chapter 16 — The Sciences

reducing the negative symptoms. With this new model, we have added some new terms to the health equation. The inner states and expectations of the patient, the caregivers, the people close to the patient and the entire community all contribute to the occurrence of disease and the course of recovery of all individuals.

Drug tests and other medical experiments

The double blind placebo-controlled drug test, which is the "gold standard" of allopathic medicine, is a wonderful example of the triumph of paradigm over observed reality. The paradigm is that only outside, physical interventions contribute to the recovery from disease. Allopathic medicine has been very unkind, even hostile, to practices that do not involve physical interventions. But it turns out that to test the effectiveness of a new drug, one has to compare the effects of taking the drug with taking a placebo. This is because of the size of the placebo effect. We know that 30 to 50% of the people who take the placebo respond positively[4]. It should be pretty obvious that physical, outside interventions are not the only influences that make disease go away, but that idea has not penetrated very deeply into current medical practice.

The quantum life model accounts for why the placebo works and why the current paradigm is not sufficient. Recovery is influenced by much more than external intervention.

The double-blind part doesn't work in the quantum life model. It is possible for individuals to know things, like whether they are getting a drug or a placebo, without any direct transfer of that information. The other purpose of the double blinding is to keep the intentions of the experimenter from influencing the patients and the outcomes. Double blinding can't do that because we now understand that influence can act across time and space and effect quantum level processes in anyone. Indeed, drugs do much better in their initial trials than in trials conducted later. The double blinding does not isolate the patients from the expectations of the experimenters and their sponsors.

There are two implications here. One is that we cannot assume that a living being is an isolated thing. All living beings are connected to all others that have some interest in them. Their health and recovery are influenced by the intentions of everyone that is involved in any way. So experiments need to be designed with that connection as a basic condition.

The other implication is that this phenomenon can be exploited to improve the treatment of dis-ease. I'll talk about that in the sections that follow.

Causes of disease

As I write this, I think it is still fair to say that the mainstream model of disease is that disease is caused by outside physical agents and internal physical malfunctions. There is ample evidence that stress, emotional problems and patterns of thought contribute, but that has not permeated much of mainstream diagnosis and treatment of the physical diseases. Now we are adding another influence to the list of non-material contributors to disease: the quantum level influence of the patient, the patient's family and friends, the caregivers and their entire community. Instead of looking outside for the causes of disease, we need to look inside.

I am not suggesting that there are no physical or outside causes of disease. I am saying that the factors that determine whether a given individual gets a disease is determined by a chain of factors which might include external agents like microbes and chemicals, but always includes internal factors like thoughts, feelings and intentions of the patient and all the people who are concerned about him or her. Note that I said "might include external agents." It is entirely possible, and quite common, for diseases to be caused entirely by internal influences.

The quantum life model adds another non-material influence on our state of health.

Treating disease

Most doctors and many complimentary/alternative practitioners like to believe that they are treating the ills of their patients, that their actions are "curing" the disease. This is fortunate because many patients expect the doctor or other practitioner to "make them well" and get quite irate if they are asked to actually contribute to their own recovery. So everyone should be happy. Unfortunately everyone isn't happy or healthy. In fact, medicine and health care in the US are in something of a mess right now. I believe that an underlying cause of the mess is this expectation that healing is something that someone else does to us.

We have seen that there is abundant evidence and experience that tells us that we each contribute a lot to our own health and dis-ease. Each of us contributes to the health and disease of others, as well.

The point I want to make is that people have the power to heal themselves (it is the only way healing occurs), so everyone should know how to use that power. The techniques are well known: meditation, relaxation, letting go of outcome, visualization, vocalizing (toning and chanting) and intentional motion. They could be taught in schools. Doctors' offices and hospitals could offer "healing practice" groups for their patients. For those who cannot attend a group, a person could go to their bedsides. A live person is more effective than audio or video recordings.

Self-healing practice should be the first action taken at the onset of any dis-ease.

Modern Western culture is the only culture where spiritual practice is separate from healing practice. It is possible to think that churches could teach healing practices as part of the spiritual practice of their members.

The role of mechanical procedures in health care

Allopathic medicine is characterized by two things: a focus on material and mechanical causes and treatments of disease, and the belief and expectation that allopathic medicine is a complete system of health care, i.e. your friendly, local allopathic physician can address all of your health care needs. Many doctors and many patients still believe this is true. For many others, it is changing.

We have seen that physical and mechanical interventions are less important under the new paradigm. When I say mechanical, I mean things like putting people back together after a car wreck. Where does that leave physical medicine? I see two possible roles for the current "mechanical doctors." One is to continue the current trend toward expanding mechanical allopathic medicine to include complementary and alternative practices. This leads to medicine becoming holistic, integrated medicine. Presumably, part of the expansion would include routine use of inner healing practices by all patients. To be effective, this change should include all physicians, not just primary care doctors. If allopathic medicine is to be a complete system of health care, then surgeons and other specialists should be aware of and embrace the use of inner healing practices by their patients and they should nurture a healing inner state in themselves.

If that suggestion sounds pretty unrealistic, here is another role for physical medicine. Leave it as physical medicine, but don't try to make it a complete system of health care. Let it be a specialty. There is nothing wrong with mechanical medicine when something mechanical in the body needs fixing, like setting broken bones. I am suggesting that the primary care practitioner be a holistic practitioner who directs the patients' personal healing practice and recommends specialists when needed. Andrew Weil, the well-known MD advocate for integrated medicine, summed up the relationship between mechanical medicine and other practices when he said, "If I'm in an auto accident, don't take me to an herbalist."

Training of doctors

I have proposed two kinds of roles for allopathic physicians: being more holistic or being more narrowly specialized in the overall spectrum of health care providers. In both cases, training doctors in the importance of their own inner state and that of their patients is vital. I trust everyone has encountered the arrogant, egotistical doctor, particularly among specialists. I am not suggesting that all physicians fall into that category, but there are more than a few of those types practicing medicine. The caregivers' inner feelings and expectations about their patients are broadcast to the patients and affect how the patients feel and respond to treatment.

The mark of a good healer is letting go of ego. I am afraid that this is an earthshaking proposition to many medical schools and practicing physicians. Being a medical specialist is a very prestigious position in our culture, which for most is a welcome side effect of being a high-powered specialist. That attitude is not conducive to effective healing, of any type. Actually, it may even be a pretty big change to even suggest that the physician's inner state is an important part of the physician's practice.

Maintaining an open, accepting inner state is as important as maintaining any other aspect of the physicians' technique. This inner state can be taught and there is a great deal of experience in teaching these things. Unfortunately, most of that experience is in the East. The Buddhists and Taoists have taught people to alter their inner state. That kind of teaching is, of course, very different from what happens in most Western schools. The big difference is the value placed on the inner state of the individual. Inner state is not an issue in Western medical schools, but it needs to become an important issue if physicians are going to make a positive non-material contribution to their patients' recovery and health.

Chapter 16 — The Sciences

Science in the new reality

Spirit is real and is intimately involved in every action we take. We are powerful spiritual beings (subtle, but profound, of course). All this means that the big science institutions in our culture that are responsible for dealing with living things, the biological sciences and medicine, are in for a profound change. That change will make doing science and medicine look more like spiritual practice in a Taoist monastery. I don't mean that scientists and doctors will sit around meditating all day, but practitioners of science and medicine and their patients will spend regular time in what we now call spiritual practices. They will probably come up with different names for the practices.

The result will be, I believe, a much more humane and holistic science and medicine. When doctors and scientists are attuned to their own inner states, they can be attuned to their patients' and subjects' inner states. Everyone should feel a lot better.

---

[1] Kuhn, T. *The Structure of Scientific Revolutions,* The University of Chicago Press, Chicago 1970. Pg 43-51.

[2] Tiller, W., Dibble, W., Kohane, M. *Conscious Acts of Creation.* Pavior, Walnut Creek, CA 2001.

[3] Wilbur, K. *Sex, Ecology, Spirituality: The Spirit of Evolution.* Shambhala, Boston, 2001.

[4] Benedetti, F. *Placebo Effects: Understanding the mechanisms in health and disease.* Oxford University Press, 2008.

Chapter 17
Religions

AH, RELIGION, MY old nemesis. While I have had my troubles with organized religion, the fact is that religions are important. They are our culture's and society's means of connecting with the spiritual, which I have been arguing is a profound influence in our lives. Religions thrive on differences and here I come, suggesting there is a common reality under all forms of spiritual practice (I'm hardly the first to do that), and proposing that science can tell us what god is made of. Worse than that, I have proposed that we are gods (I'm hardly the first to do that, either). These are big changes for current organized religion and for most personal beliefs. Heresy is the appropriate word in this context, but these religious heresies are no worse than the scientific heresies of the last chapter. So let's push on and see where this takes us.

God

God is the most basic, fundamental spiritual belief. It it probably the oldest. The belief is founded on a real phenomenon as we have seen in previous chapters. In most Western and many Eastern religions, the God (the specific god of each religion) is remote from the material world. God is in heaven. It is a long journey to heaven with many entrance requirements to get there and finally meet God. Large and beautiful buildings are erected to provide a place to worship the distant God. God is immortal, distant, all-powerful, all-knowing, often judgmental. It takes people who are specially trained to act as intermediaries between mere mortals and the God: clergy and priests.

The situation is different in the mystic traditions, especially in the East, where the belief is that god can be experienced directly through inner practices like meditation and chanting. That belief is nicely expressed in the Sanskrit greeting, Namaste, which means something like, "that of God in me honors and acknowledges that of God in you." This makes god a little closer, but still quite different from you and me.

The quantum god I have described in these pages is quite different. It is in the same "place" as our material world. It is not separate from it. The quantum god is all-knowing in that it is (consists of) all the information there is. Its power is limited to influencing quantum-level events, although on a huge scale, and then only in a probabilistic way. There seems to be some preference for forgiveness, acceptance, compassion and love, but it is not a strong preference and certainly not a demand.

In this model it is not just god that has changed, it is humans and all living things. Let's look at that next.

Congregants

People have different roles in different religions. It ranges from "having that of God within" as the Quakers like to say to being "sinners in the hands of an angry God" as Jonathan Edwards said in 1741. The notion that sin and unworthiness is the normal human state is common in Christianity and goes back to Augustine (4th century) and the idea of original sin[1]. In most cases the view is that people are very different from, and very far way from god. Except for the mystics who value direct experience of the divine, most religions provide guidance and interpretation of the divine for the congregants who presumably are unable to do that for themselves.

The model I have suggested here is very different (this is an understatement). People are gods. They have the same powers of knowing and influence. They are immortal. They live in the same place as all the other immortal and divine beings. The only differences between humans and the bigger gods is that we have material bodies once in a while and our scale of awareness is smaller.

While this is a big change, it does not mean that we should all go out and get robes and a throne and spend our days pontificating. Recall that we exercise our greatest power through forgiveness, acceptance, compassion and love. It's hard to be compassionate when you are sitting on a throne dressed in fancy robes.

If the people-as-gods paradigm were to be adopted widely, it would have implications both for individuals and the churches they attend. For individuals, it represents a very different view of self. Think about the difference between hearing that you are a miserable sinner and hearing that you are a god, or at least god-like. Suppose you grew up hearing that you are a god? This would not eliminate psychological and spiritual problems, but I believe it would reduce them considerably.

Being a god carries more responsibility than being an unworthy follower of a powerful and distant god. Unfortunately that responsibility takes the form of concern about one's inner state. Right now the fraction of the population that is willing to look at its own inner state is very small (that's based on my own observations). We don't like to look inside of us because it is filled with so much bad and disgusting stuff. I am slightly hopeful that adopting the people-as-gods model will make it easier to look inside. If we are gods, and the big god is not so different from us, then we don't have to hide our unsavory parts from the old, critical and distant god.

In the current models of god and religion, God can know our sins. In the new model, it is worse than that. The unacceptable stuff we try to hide in our shadows is exactly what we are broadcasting to the universe. We are shouting our shadow from the top of the mountain for all to hear. Even without the quantum god model, our shadows are visible to all because it is our shadow feelings and beliefs that shape our behavior and the patterns of outcomes in our lives.

Perhaps believing that we are god-like and that god is like us will make it easier to look inside and do what we need to do to become whole. That's another of those practice issues that is mentioned in the appendix but will have to wait for the next book for a full discussion.

If everyone is god then, perhaps it will be harder to condemn, dismiss and exploit others. The old reasoning went something like this: God is far away, but I know he is on my side. Since God is on my side, he must not be on yours. Now everyone else is as much a god as I am. This does not mean that all behavior is equally "of god."

Chapter 17 — Religions

The golden rule, compassion and justice remain as standards for acceptable behavior. It might mean that people who display unacceptable behavior, from rudeness to mass murder, might be treated as gods gone astray rather than alien beings of no value.

Such a change in the people who attend churches would presumably require a change in churches. Even if the existing churches chose not to change, I expect people would move to churches that did a better job of supporting their current god-like state. Let's consider the nature of these new churches next.

Churches and clergy

The nominal job of churches of all types is to serve as a channel between the religion's model of god and the religion's model of people. I have proposed models of god and people that brings them (god and people) much closer together. What would a church look like that supported this new paradigm? It would be less about control and guarding the gates of heaven and more about helping people become whole and claim their godhood.

Common ground

If god and spirit are real and live in the quantum field then there is only one spiritual domain. This domain must be shared by all organizations and practitioners seeking to know god. In other words, it provides the common ground from which all those organizations and practices sprouted. I understand that acknowledging common ground between organizations that depend on their uniqueness for their existence is unlikely. But just suppose it was possible.

What does it mean to religious organizations to say that they are all based on the same underlying phenomena? I hope it might help undermine the emphasis placed on difference in form and practice. All of the forms and practices are nominally attempts at expressing the underlying "truth" and here is the generic, non-specific "truth." I have suggested the the underlying truth is not the forms. Even the information we get from the intelligent field is abstract. All of the forms we see in religious organizations are human additions to make the underlying formlessness available to their members. This is a very useful service for a spiritual organization to provide to its members.

The world would be a calmer place if religions acknowledged that they were each a particular view on the same underlying spirit. This seems like a simple and very important idea. There will be difficulties with implementing it as long as the most important thing about a religious organization is that it is different from others and, therefore, true. I am slightly hopeful that if science were to embrace that underlying spirit, it might make the idea more appealing to people and perhaps even to their religious organizations.

Differences in religions are necessary. Most people are more comfortable with specific, literal forms in their religion. A few others are not comfortable with specific forms. The specific forms of a religion need to be appropriate for the culture. There is room for lots of variation on top of the common ground. If people could acknowledge that their specific forms were just tools to

Chapter 17 — Religions

help them to understand the underlying and universal mystery, we could save a lot of bloodshed and misery.

Does that mean that all religions are equally "OK?" No, it does not. The behavior of individuals, and collectives of individuals, like religions, are guided by the same sort of probability curves that govern all other quantum level behavior. Everything that can happen, does happen, and that includes good behavior by organized religions and perfectly reprehensible behavior by religions. The criteria to determine which is good and which is reprehensible are the values taught in the core of all religions: justice, compassion and "Do unto others as you would have others do unto you." Of course, most religions ignore those criteria in their dealings with other religions and their own members who break their specific rules.

Spiritual practice

Spiritual practice is a funny term. In the religions in my culture, it is associated with esoteric, Eastern practices like meditation and chanting in foreign languages. People here may "go to church," pray or read the bible, but those are not usually called spiritual practices. I never heard the term, spiritual practice, applied to going to church. I would like to use the term in the most general way: Whatever we do to address our relationship with the non-material aspects of our realty is our spiritual practice. So, going to church every third Sunday and serving on the building committee is a spiritual practice.

The question: Does the quantum life model have any implications for spiritual practice? I think it does. An implicit part of the spiritual model of many people in conventional religions in the West is that our connections with the spiritual are limited, isolated events, consisting of our time in church or explicit prayer. This is not universal, of course, but I believe it is a widespread idea. Another implicit feature of many religions is an emphasis on the outer life of people, that is, the words and behaviors. It is enough to participate in the observances of the religion and behave according to the standards. No inquiry is made about the inner state of the congregant.

I understand that this is the arrangement preferred by the great majority of people. Churches that only ask for outward compliance with their standards have vastly more attendees than organizations that explore the inner state. The quantum life model tells us that we are in constant, two-way communication with the non-material, creative intelligence. Indeed, we humans are creative intelligences that are continuously contributing to the unfolding of reality. Here's the rub: What we are contributing to the creation of reality is precisely our inner state, not our good words or our good deeds or even our good, conscious thoughts. Our inner state is pretty important and deserves some attention.

Now, the question is, what is the "right" spiritual practice? The answer, at least in Western culture, is that it is really hard to tell. The problem here in the West is that we separated spiritual practice from healing practice a long time ago, which leaves us with a lot of practices that deal only with the non-material aspects of our lives and this makes it difficult to evaluate them for effectiveness.

In all other cultures in the world, healing practice and spiritual practice are the same thing, so we have some basis for determining which kinds of spiritual practice work and which are not effective: We can see improvements in our physical, as well as mental, well-being when the spiritual practices are working.

My favorite practices are the yogas from Vedic India and the Taoist practices from China. The message of these practices is that being "healthy" requires addressing the health of body, mind, emotion, spirit, relationships and actions.

The goal of spiritual practice is wholeness, which is an interesting word. It is possible to be quite whole without being entirely free of problems, negative symptoms and a variety of neuroses. I am not suggesting that any kind of perfection or unblemished state is required. Wholeness implies a kind of relationship with what is.

Fortunately, if you were to do practices in all those areas (body, mind, emotion, spirit, relationships and actions), it is quite likely that you would have fewer negative symptoms and neuroses than if you avoid those practices. The usual paradox applies here: The best way to achieve a particular goal is to not be concerned about the goal and to just do the work.

My suggestion is that we focus on spiritual practices that contribute to our physical, mental and/or emotional health, which brings us to our shadow.

Shadow work

I have put a lot of words into making the point that our shadow feelings are a very important, if negative, part of our creative contribution to the cosmos. Any discussion of spiritual practice has to include shadow work. So here it is.

Assuming we want to make a positive contribution to the ongoing creation of the cosmos, shadow work is the most important kind of spiritual practice. Under my definition of the "right" spiritual practices, shadow work is a good practice because it has observable results: We can feel the change. Another was to say that is bringing our shadow feelings up to awareness and coming to terms with them is the main task in becoming whole. It also reduces our aberrant behaviors and our neuroses.

Many churches contribute to the development of shadow feelings with their emphasis on being positive. If you are a member of the right church then you are blessed and should be happy. Being righteous means you don't have "bad" feelings, like lust, jealousy, depression and the rest. Since those are all normal human emotions, they do come up and they have to be buried in the shadow to keep them safely out of sight.

I'll repeat myself here because I think it is important. A belief system that makes us gods could go a long way to making it easier to deal with our own shadows.

Power vs. subtlety

Power is an important selling point in many religions. This is because power is very popular. It always has been. Gods were the superheroes of antiquity. They had their foibles, but when they wanted to do things, Socko! Bang! They got it done. Flash forward 3,000 years and we find that action movies, superhero comic books and Harry Potter are still full of people with foibles who can exert great power over things when they put their minds to it. Great power remains very popular.

Chapter 17 — Religions

Religions have always used images of great power to attract and retain followers. The gods are described as being powerful influences in our material word and the power to make things happen is promised to people who join the ranks of the faithful and follow the rules.

The problem is that actual evidence for god or an individual wielding lots of non-material power over the material world is very scarce. In spite of that, it remains an effective tool for attracting people to books, movies and religions, which tells us it holds a strong attraction for people across many cultures.

What does the quantum life model have to say about power? Both gods and mere mortals can exert influence over events in the material world, and they all do, continuously. So the quantum life model provides a basis for the popular descriptions of heroes and gods. The difference is that the power is exercised at the level of quantum events. That is to say, the influence is subtle. Subtle? In my dictionary it is defined as delicate, deft or ingenious, not open or direct, delicate suggestive: not grossly obvious, not easily detected.

I think a clear implication of the quantum life model is that we need to cultivate a sensitivity to subtlety in our own actions and in the world around us, and to develop an appreciation of subtlety in our gods.

A circular road

I started out a long time ago, running away from religion and looking for spirit in science. The road led though fringe science and quantum mechanics. Now I have assembled an understanding of spirit in scientific terms and I find myself back at religion.

I ran away because it looked like the real purpose of religions was to exert power and control over people and culture. Now I've found what I believe to be the reality that underlies the spiritual beliefs of so many people and it is not about power and control. It is about my being whole and subtle. There are no absolutes and almost no rules, just a slight preference for goodness over evil. It does ask that I work on my shadow, but my shadow is just stuff that gets in the way of being the god I am capable of being.

I would like a religion based on those ideas, if it wasn't too organized. The mystics are pretty close. But, everyone is not me. Everyone is not even very much like me in their preferences for spiritual connection. Variety of religious experience is essential.

What can we take away from this discussion of religion in the light of quantum spirituality? For those who are seeking a way to come to terms with their religious experience, I hope this offers some food for thought, perhaps even some useful direction.

"Those who are seeking a way…" is the subject of the next chapter. I like teaching. The next chapter probably reflects my attempts to solve other peoples' problems. I have come far enough on my journey on the road to understanding that I know the only problems I can see and comment on are my own. We teach what we need to learn.

---

[1] Fox, M. *Original Blessing*, Bear & Co., Santa Fe, NM 1983. Pg 11.

Chapter 18
People

PEOPLE HAVE PARADIGMS, too. Like changing paradigms in organizations or religions, changing paradigms is hard.

Here are four groups of people who have different, though possibly overlapping, paradigms. The quantum life model has implications for them all. The four groups I want to consider are Newtonians, practitioners of complementary techniques, people who live in their heads and people trying to make sense of spirit. It may not surprise you very much to learn that I consider myself a card-carrying member in good standing of all of those groups.

I have noticed that when I look at other people, I can really only see things in them that I have in me. So when I analyze them, or offer advice, I am really offering advice for myself. I hope that you are enough like me to find some things of value in that discussion.

Confirmed Newtonian materialists

There is a considerable number of people whose main paradigm is Newtonian materialism as practiced at the end of the 19th century. The quantum revolution in physics can be isolated from day-to-day life, so it is possible to maintain the hundred-year old paradigm. Some members of this group are quite outspoken in their defense of the paradigm. They populate organizations like Quack Watch and organizations with the word "skeptic" in their title. From what I have seen, their purpose is to deny any work that conflicts with the paradigm of scientific materialism. These days, there is certainly no lack of things that need denying.

Others in these groups include a lot of traditional guy types who are uncomfortable with anything that is not concrete and material. For a traditional guy, that includes all the non-material science stuff plus emotions and feelings.

There is a shrinking, but still large, group of physicians in this group. I recall being invited to talk about the science of energy healing to a group of family practice residents. These were all recent graduates from medical school. I didn't get to the energy healing stuff because that was too "far out" for them. They were willing to talk about, and criticize, herbal remedies. At least with herbs there was a substance involved, but the very idea of healing with no material influences was too far outside their paradigm.

My assumption when I first started assembling materials to talk about these topics was that the audience would be science types who would welcome a science-based explanation of phenomena that currently do not have science-based explanations. That was a naive assumption on my part. Science people, just like everyone else, assume that their current paradigms are reality. Any threat to those paradigms is taken as a threat to their reality and so they do not come to my classes.

Chapter 18 — People

The fact is that most people take their paradigms with them to the grave. This is true even while the paradigms they adhere to are fading away in the mainstream of their profession or culture. I am reminded of my father's freshman chemistry professor (tenured) at the University of Wisconsin in 1940, who did not believe in atoms.

The quantum life model is certainly a paradigm change for Newtonians. If people are unwilling to consider things outside their current paradigm, then a new paradigm has no effect on them. On the other hand, if they are willing to consider a new paradigm, then this model offers a science-based view that can eliminate many of the discontinuities that Newtonians have to maintain. The discontinuities are all of the non-Newtonian phenomena which have to be denied or ignored in order to believe that Newton's materialism is a sufficient explanation for the world. It takes a lot of energy to deny all that. For those aware that there is more to the universe than Newton allowed, here's a pleasantly scientific approach.

It means they can maintain their scientific position and allow for the reality of those pesky healers, intuitives and psychics. There is no requirement to join a religion or anything like that.

From the point of view of the scientific observer, understanding the underlying mechanism for a phenomenon can give you a sense of ownership or control over the phenomena. I know that understanding actually confers no control, but it still feels good. I don't have to spend all the energy trying to explain why energy healing can't work to people who obviously believe that it does work.

For the Newtonians, the quantum life model lets them participate in a wider world. It reduces the number of enemies they have to confront.

Practitioners of complementary arts

Contrary to my initial assumptions about the audience for my ideas, most of the people who have attended my presentations have been people already living in the new paradigm. They are energy healers, intuitives, energy-based massage therapists, dream workers hearing the voice of god in their dreams, members of churches and people who talk to angels and spirit guides. In other words, they are people who are decidedly not Newtonian materialists.

Even though these folks have moved into the new paradigm, none of them seem to have any trouble with Newton's material reality. They don't walk into walls or cross the street without looking both ways. As a result, there is less of reality that they have to deny, compared to the Newtonians.

Nevertheless, many of these New Age practitioners still have a bit of dualism to maintain: They have to be a little careful about who they talk to. For example, nurses who practice energy healing have to be careful not to talk about diagnoses and cures with many doctors (the Newtonian materialists). Massage therapists who view their work in energy terms have to tailor their words to the state of each client. Some clients don't want to hear about energy releases or blockages.

The quantum life model gives the practitioners of the non-material arts an understanding of how their work fits into the world of material science, or material sciences extended by quantum life science. They don't have to feel that they are doing something "weird" when they are around materialists. I think I once harbored the idea that these non-material practitioners could explain what they do to a materialist so the materialist would understand them. I now know that is wishful thinking. People don't change paradigms because of a nice explanation of a new paradigm. So I am happy to be able to help practitioners feel more comfortable in their skins.

At unguarded moments, I tend to think that disciplines outside the scientific mainstream have more trouble with unfounded claims and exaggerated benefits than the disciplines in the scientific mainstream, like allopathic medicine. Just take a look through any enthusiastic nutritional or healing-type publication and you will see an amazing number of claims to cure all kinds of aliments with just a few supplements or some treatments. The quantum life model tells us that claims that imply deterministic cure, as in "take this supplement or treatment and you will be cured," are exaggerations. Health is a quantum process and so the outcomes are always probabilistic: Some people will get better and some will not. I also understand that no one would buy a product that advertised, "This product might make you feel better."

I said "unguarded moments" because when I think about it everyone, including allopathic physicians, makes the the same sort of exaggerated claims. I am thinking of ads for cancer care facilities that use chemotherapy or pharmaceutical ads. The quantum life model gives us some sanity checks on claims for healing outcomes.

The model tells us that action at a distance, or without any physical intervention, is possible. It also tells us that the particular technique is not very important. What is important is open acceptance of self by the healers so they can extend acceptance and compassion to their patients, which is the medium of non-physical healing.

People who live in their heads

"People who live in their heads" has a strange sound about it, but I assure you that it is a very real segment of our population. I put this section in I because I am one of those head people. It has taken me many years, including the effort of writing this book, to recognize the condition as a limitation and to begin to grow past, or out of, that condition.

For those of you who don't recognize the condition, I'll describe it. Head people are intellectuals, idea people, readers and thinkers. Those are all admirable attributes, of course. To qualify as a head person, you also have to ignore or repress the other aspects of your humanness: emotions, feelings, body sensations and the body in general. Logic and reason are the guiding principles for head people. We pride ourselves on living ordered, rational lives. No uncontrolled exuberance allowed here.

Western culture has a long history of emphasizing reason over the other ways of knowing. The Greeks placed great stock in reason. Kepler and Galileo showed us how to use reason to design experiments that could confirm the correctness of our reasoning. Newton gave us laws that could account for everything. There was no need for god. Emotions and intuitions were just static in the brain waves and they had no value in dealing with

Chapter 18 — People

the world. Only flighty women and young children engaged in such things, anyway.

By the end of the 19th century, reason reigned supreme. "The physics of the future will take place in the fifth decimal place." Everything was figured out. That attitude did not survive the first decade of the 20th century due to the arrival of relativity and quantum mechanics. But now, 100 years later, the value placed on reason remains unchanged. There are lots of doctors, academics, business managers, scientists and engineers and "regular guys" who wouldn't be caught dead acting on the basis of mere emotion.

What does the quantum life model mean for us head people? I will explain what it means for me and hope the head people among my readers will find it interesting.

I was raised in a family that valued intellectual achievement. My parents were dedicated to their three sons, but they were not emotive people. Feelings and emotions were never discussed or expressed. I grew up with the belief that my value depended entirely on my mental accomplishments.

I remember returning home for a visit as an adult. I had recently recovered from a bout with shingles. I mentioned this to my mother who replied, "Doug, how could you? That's nerves!" I spent most of my life trying to avoid feeling, expressing or discussing emotions and feelings. Looking back, I think I might actually be a fairly sensitive sort of person, but I'm still "thinking" about it and having a hard time doing, being and feeling.

There was one crack in the wall, however. Somehow, even after rejecting religion and adopting what I called scientific materialism, I still believed that there was something more to the universe than matter, energy and the laws of physics. Of course, I went looking for that something in science and science fiction. It took a long time, but now I think I have finally found what I was looking for in the ideas I have presented in this book.

It was quite a shock to my stoic, rational ego when I realized that the means of connecting to the intelligence of the universe was all of those feelings, emotions and felt senses that I spent a lifetime ignoring and repressing. Now I have determined that I don't have to be a fluttery, weepy pile of emotional reactions to be a feeling person. The implication is that to be able to tap into the infinite wisdom and information in the field, I have to be sensitive to subtle feelings and images. In order to have a positive influence on the things that come into my life and to be able to react positively to those events, I have to be aware of my inner state of feelings and emotions and to be able to align my conscious, rational thoughts with my inner feeling state.

On one hand, this seems terribly obvious. It is very old wisdom that to be healthy I have to be whole, which means embracing all aspects of my being. I think that stuff is important. Nonetheless, it was a surprise to me when I realized that to participate in this "new" world, I would have to embrace things that I had avoided for most of my life. Worse than that, I found that much of my behavior, far from being directed by my careful, logical, rational thoughts, was directed by subtle, but compelling feelings that I was only vaguely aware of. There were situations that I avoided and things that I did not do. Many of the things I avoided would have been very helpful to me.

Chapter 18 — People

The real eye-opener for me came when I realized that it was not easy to change behaviors that are driven by long established emotional and body feelings. Me! The eminently rational, logical, in-control guy was being led around by feelings and emotions just like all the people I had cluck-clucked about all my life. You know, the ones who keep doing things that are obviously hurting them and they just keep on repeating the same behavior, with the same, unsatisfactory results. It did relieve me of the burden of being better than most people I see, but it was still a nasty shock.

In the past, I was aware that being the way I was might not be the optimum state of health. I was not entirely whole, but I viewed it as a character flaw, a subject of endless introspection and self help efforts. With this new information, I now see a whole new level of implication. Being disconnected from my feelings and emotions is bigger than just my character. It defines the influence I have in the wider world. While I'm not conscious of my inner feelings, they are probably not entirely healthy. In fact, I now know that they are what drives me to isolate my feelings from myself and from everyone else. They are also what I broadcast to the universe. They are my subconscious, continuous prayer to the powers of the universe. I don't really want to be doing that sort of thing.

But change is hard, in spite of all the promises made by the self-help and personal transformation books. I see one of the major applications of this new knowledge is to provide additional incentives for making effective change in my life, and in the lives of all those head people out there. It does not seem to me to be more than that. This new knowledge is just more knowledge. It doesn't directly make me, or anyone else, actually do anything different. I am hopeful that because it is knowledge from science that it will serve as a better motivator for me and my fellow head people.

People who are trying to make sense of spirit, religion and reality

I assume that I am not unlike lots of other people out there. I am interested in learning about my world and environment. There is something in the idea of spirit that is very appealing, but much of what I see in organized religions was clearly fabricated for very non-spiritual ends. This makes it hard to go to church to meet my spiritual needs. The quantum life model provides a view of reality and spirit that I can live with.

There is intelligence "out there." We can communicate with it. It is the source of all wisdom, invention and insight. The "thinking" of that universal intelligence is action in the material world that is both profound and subtle. The intelligence is literally the ground of all being. I am part of "all being," so I am part of the intelligence of the universe. I can tap into that intelligence if I can manage to quiet my conscious chattering enough. My thoughts (feelings and emotions, really) are actions in the material world that are both subtle and profound.

What does all that mean when I am trying to decide what to do Sunday morning? I am still free, of course, to do whatever I want. But now I have a context to understand what I am doing. Organized religions are, at their best, attempts to provide a concrete, culturally appropriate physical interpretation of the abstract and formless ground of all being. So if I go to a church or temple or whatever that matches my cultural and social sensibilities, I can participate in the interpretation offered knowing that it is representing fundamental truths about the nature of the universe.

If the particular church I choose also engages in practices that are clearly not expressing truths about the ground of all being, like accumulating economic, social and political power in the hands of the clergy, not treating others as they would be treated, repressing their own membership (women, for example) or people outside their membership, or selling the disease to make it easier to sell the cure (like branding sex as dirty and sinful), then I can address the injustice, ignore it or leave, knowing that the underlying truths are intact. If I leave, I can know that I am not leaving the source of truth, but just one interpretation of the underlying reality.

What of the other great religion in our culture, science? Science, of course, does not like to think of itself as a religion, but it is. Orthodox science expects the same level of faith and denial of things outside the faith as the most orthodox and conservative religions. Science does have a better record of allowing the occasional change in paradigm, but the changes are always messy affairs.

The quantum life model tells us that Newton, Descartes and Kant got the big matter stuff right, but they completely missed out on the non-material aspects of reality. God and spirit live in the ground of our material reality: the quantum field. The basic processes of life are quantum-level processes that continuously exchange information (energy really) with the distributed and intelligent quantum field.

So it is entirely possible to sit in a science paradigm and be comfortable with a large, non-material component of reality. Of course, it would be a very different science paradigm from today's mainstream science paradigms. While you are not likely to get killed for changing paradigms, you can easily get passed over for promotions, have your papers rejected for publication or get fired.

The implication for those of you who are uncomfortable with the inconsistencies between science and religion, between material and spiritual, is that it is possible to live comfortably with both. Of course, that requires a different paradigm from both current science and current religion.

Being one, being god

The universe is made of intelligence that favors connection and compassion in a probabilistic sort of way. The living things with material bodies think and feel with the same intelligence as the rest of the universe. Our connection to the universe and to other living things depends on our inner state being one of acceptance and compassion. Connection, in this model, is the same as influence. Our ability to influence others in a positive way depends on the same inner state of compassion and acceptance. Our physical, mental, emotional and spiritual health are all the same thing. Physical and mechanical medicine has a place, but it is not the foundation of health and healing.

Being god-like is being aware, accepting and compassionate toward your self. Then you can connect with others and be a healing presence for them. Then you can connect with the larger intelligences that permeate our universe and be aware of them. Connection is also influence in this world. The influence exerts itself in quantum level processes. The quantum level processes that we are concerned with here are the decisions, behavior and health of ourselves and our fellow humans.

When we are being one, we are connected with all that we perceive, the material and the non-material. When people in this state come together to form organizations, those organizations will be aware, accepting and compassionate toward their members and those outside the organization.

It is not that this is new information. It has been taught for as long as sages have taught their fellow beings about right living. What is new is that materialist science is now discovering the non-material world that it has ignored for the last 300 years. If we are a culture enthralled with science, now perhaps science can move us toward embracing our real identity, being one with god.

Appendix 1
The Equations

Fourier Transforms

The Fourier transform is a translation of a function in normal space, three spatial dimensions and time, to a function in frequency space, inverse time (frequency) and inverse distance. If the normal space function is f(x), where x is one or more of the normal space dimensions, then its Fourier transform is g(r), where r is one or more of the inverse dimensions.

The equation defining the Fourier transform, g(r) of normal space function, f(x) is

$$g(r) = (2\pi)^{-1/2} \int_{-\infty}^{\infty} f(x) e^{-ixr} dx$$

If we have a Fourier transform, g(r), we can calculate the normal space function, f(x) by taking the inverse transform,

$$f(x) = (2\pi)^{-1/2} \int_{-\infty}^{\infty} g(r) e^{ixr} dr$$

The only difference between the transform and the inverse is the sign on the exponent of e. e is the base of the natural logarithms.

Life and Spirit in the Quantum Field

Schrodinger's Equation

Schrodinger's equation describes the movement of a single particle of mass, m, in time. That is, how the position varies over time.

The equation is

$$i\hbar \frac{\partial \psi}{\partial t} = \frac{\hbar^2}{2m} \nabla^2 \psi + V\psi$$

i is the imaginary number

ψ is the wave function. For our purposes it is the quantum field

$\hbar$ Is Planck's constant

m is the mass of the particle

V is the field that the particle is moving through

∇^2 is the gradient of the wave function. It describes how fast the wave function changes over space

Schrodinger's equation tells us that how the wave function changes with time is equal to the rate of change of the function in space plus the force exerted by the field. The solution to the equation is the wave function. It gives the probability curves for the particle's location as a function of time.

Appendix 1 — The Equations

Here is a wave function for a particle with mass

$$\psi(x^a) = e^{-i P_a x^a / \hbar}$$

This equation tells us that ψ is a function of spatial dimensions, x^a, and that it is equal to e raised to a power that is -i times the momentum of the particle, P_a, times the dimensions, x^a, all divided by Planck's constant. Notice the similarity of this wave function with the exponential factor in the Fourier transform equations on the previous page.

Appendix 2
What Do I Do Now?

This book is full of new skills to learn and new ways of being (depending, of course, on where your are starting). The way to learn those new skills is older than people: practice. Being a person who likes to start new practices, I felt some obligation to include some practice suggestions in the book, even though the book is decidedly a book of information. I understand that information seekers, like myself, don't usually take up new practices based on what they learn. (That's why I'm bigger on starting practices than keeping them) That is why this section on practices is short and in the appendix. Doing justice to the subject of practice in quantum life is another book.

Practices

I have said that the practices needed to be one with god and to be god-like are old. They have been taught by sages for as long as there have been sages, so I don't have to invent anything. Instead, I can refer the interested reader to people who have done an excellent job describing and teaching those practices.

Of all the practice sources that I know about, I think Ken Wilber has done the best job of assembling a complete and coherent set of practices. For those who don't know Wilber, he is a philosopher of human development. In his many books, Wilber has assembled all religions, healing practices and development models into one giant synthesis in his book, *Sex, Ecology, Spirituality: The Spirit of Evolution*[1]. From this he has proposed the integral models of business, education, health and spirituality. He has produced a book called *Integral Life Practice*[2]. It is the best practice guide I have seen because one of the four core modules of practice is devoted to working with your shadow.

The shadow is the set of subconscious feelings you have about your self and your place in the world. This is what we are very effectively broadcasting to the universe. It is our contribution to the making of reality. For most of us, our shadows are not contributing very much positive energy to the world. It is unique to have shadow work figure prominently in a life practice.

The four core modules are: body, mind, spirit and shadow. There are additional modules for ethics, work, relationships, creativity and soul. Each module suggests several practices with one practice recommended, the "Gold Star" practice. No time? There are recommendations for a one-minute version of each practice.

The problem with "how-to"

Supposing you like these suggestions and you run out and buy the book. You take it home and intend to really get into the practice. Then what happens? If you are like me and lots of other people here in Self-Help Nation, you don't do the practice for very long. I know there are lots of people who do that because of all the bread machines, juicers and exercise equipment I see in the re-sale shops. These are things that people bought with good intentions to use and those spiffy machines just sat around taking up space. Then there are all the self-help books sold. For the millions of copies bought and read, there has not been a corresponding surge in skinny, happy people.

I have said before that our behavior is controlled by our inner feelings. We can consciously think we want to do something, we can have the how-to information, but we won't do it if it is not consistent with our inner feelings, with our shadow. Most of the self-help books provide lots of how-to information. My observation from reading my share is that, for the most part, the how-to information is pretty good. That is, if we do what the book says, we have a pretty good chance of getting the promised results. If we don't do the work, then we need some shadow work.

Help for the Shadow

The shadow practice in Wilber's book is similar to practices in other books. It is a fine practice for working by yourself. For those who need industrial strength shadow work, I have a couple of suggestions. One is NeuroLinguistic Programming, or NLP. The practice explicitly recognizes that there are limiting beliefs that interfere with doing what we say we want to do. It deals with those limiting beliefs using all the senses and all the body sensations. There are practitioners you can go to, classes you can take online or in person and lots of books. The NLP book I have on my shelf is *Beliefs* by Robert Dilts, Tim Hallbom and Suzi Smith[3].

My second suggestion is *Healing the Shame That Binds You* by John Bradshaw[4]. In his model the reason we put things in our shadows is that we were shamed for having or expressing those things. He describes several techniques for dealing with the shame and the stuff in our shadows, including NLP. Some of his experience, which he shares in the book, is very close to my experience. I recommend it.

Of course, professional help is available if the do-it-yourself approach does not work. If you go that route, I suggest that you include some somatic (body-centered) practitioners. It is too easy for the verbal, conscious mind to protect the inner state that you are trying to change when talk therapy is being used.

A shadow work practice

I have gone to some length to stress the importance of shadow work. For those readers who think that might be true, but who don't want to go get another book, I will describe a brief shadow work practice here. This is a variation on the awareness and acceptance practices described in the books I have mentioned. Once this practice is set up, you can use it anywhere, any time you become aware of a pending shadow event.

The purpose of this practice is to acknowledge and accept our current way of being and to learn a new way to respond to a situation that has been governed by your shadow.

Preparation

Pick a situation that is driven by your shadow feelings. The easiest way to do that is to look for recurrent patterns and outcomes in your life that are not what you want. Pick one situation you are trying to change. Go through the situation in your sensory memory: who is

Appendix 2 — What Do I Do Now?

there, what do they do and say, what do you do and say, how do you feel before, during and after the situation.

Here's the part that I found helpful. Pause and become aware of who the audience is. When things happen that you don't like there is usually an audience. This is a critical or judgmental person or people. It could be a real person, like your parents or a teacher from grade school. It could also be a generalization of a real person, like your inner critic. Call this the old outcome.

Now imagine what you would have to do differently to make the situation come out more the way you want it to be. What skill or resource do you need? Leave everyone else in the situation as they were.

Pick a new audience. Imagine someone who is accepting and supportive of you. This can be a real person from another part of your life, or it can be someone you make up to be accepting and supportive.

Run through the situation in your mind using the new resource. Try it several times with adjustments until the outcome is what you want it to be. Let your actions change and imagine the other people reacting realistically. Be aware of your new audience. Feel their support. Notice how you feel when the situation turns out better. Call this the new outcome.

Finally, pick a safe, comfortable place to rest. This can be a real place where you like to be or it can be an imaginary place.

The practice

Now you're ready to do the practice.

1) Run through the situation with the old outcome until the feelings about the situation are strong and they feel real.

2) Go to your safe place in your imagination. Get into the feeling of comfort and safety.

3) Run through the situation with the new outcome. Repeat it as often as you need to have a strong feeling about the new outcome. Fill in as much sensory detail as you can. See and feel the new audience supporting and approving of what you are doing.

4) Go back to your safe place to return to the real world.

Applying the practice

You can do this practice any time you want to work on learning the new skill. An especially good time to do the practice is when the target situation is imminent. The best time is just before you enter the situation. That requires that you be aware that the situation is about to begin. If you can pause in the middle, that's good. Just after the situation is good too. If you can't catch yourself in the situation, you can imagine the last time it occurred.

Play the situation with the new outcome until you are comfortable with the new feelings. Here's the important step: Then try taking the new actions when the situation comes up again. Don't be hard on yourself if you don't produce the new outcome on your first or second try. Stay with the practice until it starts to move.

The ultimate practice

I tend to be goal oriented. I'm always trying to be better or do better. The practices I gravitate toward usually have some specific goal or outcome. These practices are OK. They can help you achieve the results you want, if you do them, of course. But I have learned a few things in the explorations that led to this book. One of them is that the ultimate practice is awareness and acceptance of self. The basic form of this practice is meditation, in any of its myriad forms. Wilber's spirit module is a good place to start.

Another way to describe "the ultimate practice" is this: This ultimate practice is the one you are actually willing to do regularly. Almost anything is better than nothing. Almost anything, subject to the caution that follows.

A caution to head people

I have known people who were pretty serious practitioners of meditation or healing arts. They took classes and they practiced regularly. They also did not appear to have benefited from all that work. What's wrong? They were not letting the practice in. They were not letting the practice past their conscious minds. I know from personal experience that it is entirely possible to do everything from meditation to self-help to psychotherapy and still keep meaningful change at arm's length. I can insulate myself from whatever the practice is trying to achieve.

For those of you who resonate with that description, I would like to augment Wilber's suggestions with two of my own: The first is that you cultivate body awareness and the next is that you cultivate feeling and emotion awareness in your body. They are almost the same thing, but not quite. I think it is helpful to start out with just physical sensations before adding the feelings and emotions.

Body Awareness

Wilber's Integral Life Practice has a body module. It addresses all three bodies: gross (physical), subtle (energy) and causal (being). If you do all three, it will increase your body awareness. When I said that head people should cultivate body awareness, I was referring to the subtle body awareness in Wilber's terms. The practices for exercising the subtle body include tai chi, qi gong, breathing practices and visualization. For serious head people, I suggest that you start by relaxing and breathing: be aware of your breath, the sounds, the feelings of the air moving through your body, the feelings of your chest and abdomen moving with each breath. Then, while you continue breathing, expand your awareness to the rest of your body. Let your mind's eye scan slowly over your body. As your attention moves to each new area, move the muscles and joints in that area very slightly, just enough to feel the movement. Notice how the muscles feel. Notice how the slight movement feels.

The purpose of this exercise is to become aware of your body and the physical sensations that live there. For a head person, this can be a very interesting experience.

Feeling and emotion awareness

After you are comfortable with how you feel physically, I suggest that you expand your awareness to your feelings, as in, "I feel bad ... mad ...happy ...relieved, etc."

To do that, repeat the opening of the previous exercise: Sit, relax, breathe. Then bring to mind a recent situation where you had some reaction, good or bad. Just sit in the memory of that situation and see how your body is feeling. Then repeat the scan and slight movements of the previous exercise while trying to hold the memory and feeling of the situation. Look for the places in your body where the feeling from the situation resides. When you find it, just sit with the feeling and the memory of the situation.

With this practice, you can become aware of information that is coming though your body. This is useful because our bodies do a better job of conveying intuitive information than our words.

The main reason for doing these sorts of exercises is so that you can be aware of your feelings when you are doing other kinds of practices. It allows you to sense whether you are fully participating in the practice or just going through the motions with your head. Developing this skill first will save lots of practice time.

With this awareness as a starting point you can move through any practice you choose with grace and persistence.

Being and doing

Being healthy, healed, whole and holy sounds so appealing. The paths and the practices that take us there are well known. The journey requires no money, fancy equipment or special spaces. There are books, classes, DVDs, teachers and groups to help us on our way. All right, those things take some money, but the basic truth is that the only real obstacle to our all being whole and holy is ourselves. The easiest person in the world to fool is yourself.

What the journey does require of us is openness, honesty and acceptance of that person in the mirror.

With those things, we can persist in the face of all the obstacles we put in our path.

So start on the journey. Be gentle and loving with yourself.

Keep taking the steps. It is a lovely journey.

---

[1] Wilbur, K, *Sex, Ecology, Spirituality: The Spirit of Evolution*, Shambhala, Boston. 2001

[2] Wilbur, K., Patten, T.,Leonard, A., Morelli, M., *Integral Life Practice*, Integral Books, Boston. 2008.

[3] Dilts, R., Hallbom, T., Smith, S., *Beliefs*, Metamorphous Press, Portland OR. 1990

[4] Bradshaw, J. *Healing the Shame That Binds You*, Health Communications, Inc., Deerfield Beach, FL. 1988

Index

A

acceptance
  as healing 285
  as power 283
action at a distance 192
afterlife
  as beliefs 257
Allopathic medicine 301
alternative practitioner 324
association
  in memory 200
atoms
  problem with 74
attractor   See: strange attractor

B

bad ideas 214
beliefs
  as afterlife 257
Benveniste, Jacques 158
big bang 267
biophotons
  as chi 167
  communication between cells 155
  emitted from DNA 155
  history 151
  in people 155
body awareness 346
Bohm, David 82
  EPR experiment 95
  Pribram collaboration 116
Bohr, Neils 47
  atom model 75
Bradshaw, John 341
buckyballs
  in two slit experiment 88

C

Carrey, Jim 286
causality
  in non-temporal experiments 97
cause and effect 296
centrioles 130
chi
  as biophotons 167
Christianity
  as heaven-centered 33
churches 313
clairvoyants 124
clockwork model 25
congregants
  view of god 310
connection
  as values 279
connections, non-local 111
  plants 111
  social animals 112
consciousness
  and light 170
  microtubes in 131

349

content addressable memory 144
   as associations 200
Copernicus 24
creativity
   process 207
curing
   vs healing 222

D

Davenas, Elizabeth 158
de Paris, Francois 235
death prayer 223
Descartes, Rene
   separating mind and body 25, 28
Dilts, Robert 341
disease
   cause of 302
   treating 303
doctors
   training of 305
Dossey, Larry 223, 242
dualism
   in reality models 55
   in religion 34
   in science 34

E

earth-centered religion 32
Eastern religions
   description of god 266
Edwards, Jonathan 310
ego
   in healers 306
Einstein, Albert
   EPR experiment 95
   papers 47
   quantizing light 75
Einstein-Podolsky-Rosen experiment 93
   non-temporal 97
electromagnetic waves
   in cell communication 161
electrons
   soul of 103
ELF radiation
   in cell communication 160
emotions
   and neurotransmitters 158
entangled particles 94
evolution 268
experiments
   effect of knowing 296

F

feelings
   and behavior 181
   and the future 247
   as action 180
   as connection to the field 186
   non-local, non-temporal 138
Feynman, Richard 281
Fourier transform
   and light 62
   as a lens 82
   in visual perception 120
   inverse 61
   of a function 60
   origin 57
   sine waves in 58
Freeman, Morgan 286
Fuller, Buckminster 88
function

Index

Fourier transform of 59
future
   and free will 242
   knowing 241
   nature of 245

G

g/God
   as distant being 309
   capitalization 2
   development 34
   goodness 216
   in mystic traditions 310
   levels of 263
   nature of 262, 264
   people as 311
   quantum god 310
   scale of 259
   ye are gods 50
Gabor, Denis
   hearing model 122
   hologram inventor 64
   Pribram collaboration 116
Galileo 24
Gilbert, Elizabeth 206
Gurwitsch, Alexander 152

H

Hallbom, Tim 341
head people 326
healing
   as acceptance 285
   as influence 184
   vs curing 221
Healing Touch 185
health 219

as quantum process 151
heaven 254
heaven-centered religion 32
Heisenberg uncertainty principle 94
holo-movement 82
hologram
   history of 64
   in the brain 126
   nature of 57
   of time 83
   properties 67
holographic memory
   content addressable 144
   in computers 143
holonomic brain 116
homeopathy
   Benveniste and 159

I

influence
   changing probability curve 186
   large scale effects 234
   nature of 179
   negative 223
   over matter 229
   quantum limitations 187
   restrictions on 176
   scale of 286
   source of 270
   subtle, strange attractor 6
information
   data vs meaning 205
   types available 201
inner experience 47
Inquisition 24

integral medicine	300
interference	
photons	87
intuitive information	
perceiving	124

J

Jansenist miracles	234

K

Kepler	24

L

life	
as a quantum process	156
as light	170
sciences	107
light	
and Fourier transforms	62
non-local	63
love and compassion	
as connection	279

M

Marias, Eugene	
termite studies	114
matter	
influencing	230
McTaggart, Lynne	98
medical intuitive	201
accessing memory	142
medicine	
implications for	299
memory	
as intuitive information	201
emotion component	200
holographic	142
not in brain	141
microtubes	
as field connector	129, 136
as superconductors	163
in synapses	137
light conductors	168
quantum behavior	131
tubulin in	129
mind over matter	26, 175
Minkowski, Herman	82
models	
as reality	13
building	13
changing	15, 38
from experience	23
inconsistent	19
levels	19
molecules of emotion	157
morphic field	49, 269

N

negative intent	
effectiveness	223
negative message	213
neural net computers	126
NeuroLinguistic Programming,	
as shadow work	341
neurons	126
microtubes in	130
neurotransmitter	157
molecules of emotion	157
Newton, Isaac	24
alchemy	28
clockwork model	25
Newtonian materialist	321
non-locality	

Index

definition 70
  in a hologram 70
non-temporal
  experiment 97

O

objective reality 293

P

paradigms
  in science 16, 294
paranormal
  widespread experience 41
Pascual-Leone, Alvaro 123
patriarchy
  defined 30
  results 30
  valued attributes 31
Penrose, Roger 93, 267
people
  as gods 311
perception
  of non-material intelligence 261
Pert, Candace 157
photo repair 153
photoelectric effect 74
physics
  anomalies in 73
  multiple equations 100
placebo
  in drug tests 301
Planck, Max 46, 74
Popp, Fritz 153
power
  as acceptance 283
  subtle power 318
Pribram, Carl
  holonomic brain model 116
  visual perception 116
Princeton Engineering Anomalies Research 231
probabilistic outcomes 77, 187
probability distribution
  in influence 178
Puthoff, Hal 203

Q

quantum biology 107
  effects 109
quantum bits
  microtubes 137
quantum field
  as Fourier transform 81
  as god 216
  as holographic memory 146
  scale in 104
  soul in 255
  summary 193
  talking to 211
  the future in 243
  thinking 208
quantum god 275
quantum intelligences
  scale of 258
quantum level
  behavior 75
  scale of quantum field 108
quantum level action 178
  nature of 179
quantum mechanics
  Copenhagen interpretation 47
  holographic field 56

single equation	100	measuring probability	189
quantum transition	132	spirit in	46, 50
quantum weirdness		training scientists	295
as reality	76	versus religion	45
experiments	85	self healing	220, 304
		shadow	
R		and behavior	228
Radin, Dean	229	and health	228
random event generator	231	as influence	228
reality		integral life practice	340
changing	11	practice	342
multiple	35	work	317
normal	11	Shchurin, S. P.	155
Taoist description	107	Sheldrake, Rupert	48
reincarnation		large-scale influence	269
Christian religion	254	morphic field	269
Eastern religions	254	plant connections	112
religion		resonance as infuence	238
earth-centered	32	sine waves	
heaven-centered	32	in hologram	58
patriarchal	31	Smith, Suzi	341
power	176	snowflake	268
versus science	45	soul	254
Remen, Rachel	283	in the field	255
remote viewing	203	space-time	82
Rogers, Carl	283	spiritual domain	
		heaven, hell	256
S		soul, ghosts	253
		spiritual entities	257
scale of awareness		spiritual practice	315
and quantum intelligence	260	spiritual seekers	330
science		spiritual values	
and spirit	297	as connection	279
Eastern	48	Stone, Ruth	206
fringe	41	strange attractor	5
implications for	293	as subtle influence	6
mainstream	41	structure and order	209

Index

super radiance
   controlling form   169
   in the body   166
superconductivity
   as quantum state   165
   in magnets   164
synapses
   holograms in   126
synchronicity   225
   as web of influence   226

T

Taoism
   as earth-centered   33
Targ, Russel   204
The Secret   183
Therapeutic Touch   185
thought
   as action   180
   as holograms   125
   non-local, non-temporal   138
   summary   194
Tiller, William   296
time
   in holograms   83
   material existence   100
truth   21
tubulin   129
two path experiment   89
two slit experiment   86

U

ultraviolet catastrophe   73

V

visual perception   116
   Fourier transforms in   120

W

water
   retains ELF   160
wave function
   collapse of   101
Weil, Andrew   305
wholeness
   and pain   221
Wilber, Ken   300
   integral life practice   339
   shadow work   340

About the Author

Doug Bennett grew up in a family with attentive but Cartesian parents. His third generation Scottish mother and engineer father wouldn't talk about feelings if their lives depended on it. Church was for appearances. Being a first child he carefully emulated his parents: the aversion to expressing feelings stuck, but the superficial view of spirit did not. Two books were very influential to his adolescent spiritual development. One was *Shaw on Religion* and the other was *Letters from Earth* by Mark Twain. *Shaw on Religion* is a collection of George Bernard Shaw's writings on religion and sprit. *Letters from Earth* is letters purportedly written by Satan during a visit to earth reporting back to his fellow archangels about the state of creation. Both authors were eloquent in their descriptions of the inconsistencies and abuses of organized religions, but both were also deeply spiritual. They described an appealing way of relating to things beyond the material. So Doug spent his adult years being an engineer like his father (BS and MS in Chemical Engineering), avoiding church whenever possible and reading science fiction and fringe science looking for some evidence of things beyond the material world. After the death of his mother (not properly grieved) and a stressful time at headquarters of a large computer and software company he took up yoga, energy healing and music healing.

He started teaching a class on how energy healing worked. That led to the world beyond the material world and ultimately to this book, which is about people being gods and the importance of feelings. Sometimes the things you run away from are the very things you should be looking at.

He lives with his wife, Pat, of 39 years who lived across the street from him when they were four years old. He and Pat have two grown daughters.

www.ingramcontent.com/pod-product-compliance
Lightning Source LLC
Chambersburg PA
CBHW021419170526
45164CB00001B/14